METHODS IN CELL BIOLOGY

VOLUME 21

Normal Human Tissue and Cell Culture
A.   Respiratory, Cardiovascular, and
Integumentary Systems

Advisory Board

METHODS IN CELL BIOLOGY

Prepared under the Auspices of the American Society for Cell Biology

VOLUME 21
Normal Human Tissue and Cell Culture
A. Respiratory, Cardiovascular, and
Integumentary Systems

Edited by

CURTIS C. HARRIS
NATIONAL CANCER INSTITUTE
NATIONAL INSTITUTES OF HEALTH
BETHESDA, MARYLAND

BENJAMIN F. TRUMP
SCHOOL OF MEDICINE
UNIVERSITY OF MARYLAND
BALTIMORE, MARYLAND

GARY D. STONER
NATIONAL CANCER INSTITUTE
NATIONAL INSTITUTES OF HEALTH
BETHESDA, MARYLAND

1980

ACADEMIC PRESS
A Subsidiary of Harcourt Brace Jovanovich, Publishers

New York   London   Toronto   Sydney   San Francisco

ACADEMIC PRESS, INC.
111 Fifth Avenue, New York, New York 10003

United Kingdom Edition published by
ACADEMIC PRESS, INC. (LONDON) LTD.
24/28 Oval Road, London NW1 7DX

LIBRARY OF CONGRESS CATALOG CARD NUMBER: 64–14220

ISBN 0–12–564121–4

PRINTED IN THE UNITED STATES OF AMERICA

80 81 82 83   9 8 7 6 5 4 3 2 1

*We dedicate these volumes to the memory
of our friend and colleague, Dr. Douglas Janss*

CONTENTS

5. *Maintenance of Human and Rat Pulmonary Type II Cells in an Organotypic Culture System*
William H. J. Douglas, Ronald L. Sanders, and Karen R. Hitchcock

6. *The Human Alveolar Macrophage*
Gary W. Hunninghake, James E. Gadek, Susan V. Szapiel, Ira J. Strumpf,
Oichi Kawanami, Victor J. Ferrans, Brendan A. Keogh, and Ronald G. Crystal

7. *Culture of Cells and Tissues From Human Lung—An Overview*
Curtis C. Harris

8. *Culture of Aorta*
H. Paul Ehrlich

9. *Human Endothelial Cells* in Vitro
Gary E. Striker, John M. Harlan, and Stephen M. Schwartz

10. Human Arterial Wall Cells and Tissues in Culture

Richard M. Kocan, Ned S. Moss, and Earl P. Benditt

11. The Fetal Mouse Heart in Organ Culture: Maintenance of the Differentiated State

Joanne S. Ingwall, William R. Roeske, and Kern Wildenthal

12. The Cultured Heart Cell: Problems and Prospects

Melvyn Lieberman, William J. Adam, and Phyllis N. Bullock

13. Organ and Cell Culture Applied to the Cardiovascular System—An Overview

Earl P. Benditt

14. Explant Methods for Epidermal Cell Culture

Susan M. Fischer, Aurora Viaje, Gerald D. Mills, and Thomas J. Slaga

15. Serial Cultivation of Normal Human Epidermal Keratinocytes

James G. Rheinwald

16. Dermal Fibroblasts

Richard G. Ham

17. Prospects for Growing Normal Human Melanocytes in Vitro

Sidney N. Klaus

18. Integumentary System—An Overview

Stuart H. Yuspa

LIST OF CONTRIBUTORS

Numbers in parentheses indicate the pages on which the authors' contributions begin.

WILLIAM J. ADAM, Department of Physiology, Duke University Medical Center, Durham, North Carolina 27710 (187)

LUCY A. BARRETT, Department of Pathology, University of Maryland, Maryland Cancer Program, and Maryland Institute for Emergency Medicine, Baltimore, Maryland 21201 (1)

EARL P. BENDITT, Department of Pathology, School of Medicine, University of Washington, Seattle, Washington 98195 (153, 205)

KATHRYN H. BRADLEY, Pulmonary Branch and Pathology Branch, National Heart, Lung, and Blood Institute, National Institutes of Health, Bethesda, Maryland 20205 (37)

PHYLLIS N. BULLOCK, Department of Physiology, Duke University Medical Center, Durham, North Carolina 27710 (187)

RONALD G. CRYSTAL, Pulmonary Branch and Pathology Branch, National Heart, Lung, and Blood Institute, National Institutes of Health, Bethesda, Maryland 20205 (37, 95)

WILLIAM H. J. DOUGLAS, Department of Anatomy, Tufts University School of Medicine, Boston, Massachusetts 02111 (79)

H. PAUL EHRLICH, Shriners Burns Institute, Massachusetts General Hospital, Harvard Medical School, Boston, Massachusetts 02114 (117)

VICTOR J. FERRANS, Pulmonary Branch and Pathology Branch, National Heart, Lung, and Blood Institute, National Institutes of Health, Bethesda, Maryland 20205 (37, 95)

SUSAN M. FISCHER, Biology Division, Oak Ridge National Laboratory, Oak Ridge, Tennessee 37830 (207)

JEAN-MICHEL FOIDART, Laboratory of Developmental Biology and Anomalies, National Institute of Dental Research, National Institutes of Health, Bethesda, Maryland 20205 (15)

JAMES E. GADEK, Pulmonary Branch and Pathology Branch, National Heart, Lung, and Blood Institute, National Institutes of Health, Bethesda, Maryland 20205 (95)

RICHARD G. HAM, Department of Molecular, Cellular and Developmental Biology, University of Colorado, Boulder, Colorado 80309 (255)

JOHN M. HARLAN, Departments of Pathology and Medicine, University of Washington, Seattle, Washington 98195 (135)

CURTIS C. HARRIS, Human Tissue Studies Section, Laboratory of Experimental Pathology, National Cancer Institute, National Institutes of Health, Bethesda, Maryland 20205 (15, 113)

KAREN R. HITCHCOCK, Department of Anatomy, Tufts University School of Medicine, Boston, Massachusetts 02111 (79)

GARY W. HUNNINGHAKE, Pulmonary Branch and Pathology Branch, National Heart, Lung, and Blood Institute, National Institutes of Health, Bethesda, Maryland 20205 (95)

JOANNE S. INGWALL, Department of Medicine, Harvard Medical School and Peter Bent Brigham Hospital, Boston, Massachusetts 02115 (167)

YOICHI KATOH, Human Tissue Studies Section, Laboratory of Experimental Pathology, National Cancer Institute, National Institutes of Health, Bethesda, Maryland 20205 (15)

OICHI KAWANAMI, Pulmonary Branch and Pathology Branch, National Heart, Lung, and

Blood Institute, National Institutes of Health, Bethesda, Maryland 20205 (37, 95)

BRENDAN A. KEOGH, Pulmonary Branch and Pathology Branch, National Heart, Lung, and Blood Institute, National Institutes of Health, Bethesda, Maryland 20205 (95)

SIDNEY N. KLAUS, Dermatology Service, VA Medical Center, West Haven, Connecticut 06516 (277)

RICHARD M. KOCAN, Department of Pathology, School of Medicine, University of Washington, Seattle, Washington 98195 (153)

MELVYN LIEBERMAN, Department of Physiology, Duke University Medical Center, Durham, North Carolina 27710 (187)

GERALD D. MILLS, Biology Division, Oak Ridge National Laboratory, Oak Ridge, Tennessee 37830 (207)

NED S. MOSS, Department of Pathology, School of Medicine, University of Washington, Seattle, Washington 98195 (153)

GWENDOLYN A. MYERS, Human Tissue Studies Section, Laboratory of Experimental Pathology, National Cancer Institute, National Institutes of Health, Bethesda, Maryland 20205 (15)

JAMES RESAU, Department of Pathology, University of Maryland, Maryland Cancer Program, and Maryland Institute for Emergency Medicine, Baltimore, Maryland 21201 (1)

JAMES G. RHEINWALD, Laboratory of Tumor Biology, Sidney Farber Cancer Institute, and Department of Physiology, Harvard Medical School, Boston, Massachusetts 02115 (229)

WILLIAM R. ROESKE, Department of Internal Medicine, School of Medicine, University of Arizona, Tuscon, Arizona 85724 (167)

RONALD L. SANDERS, Department of Anatomy, Tufts University School of Medicine, Boston, Massachusetts 02111 (79)

STEPHEN M. SCHWARTZ, Departments of Pathology and Medicine, University of Washington, Seattle, Washington 98195 (135)

THOMAS J. SLAGA, Biology Division, Oak Ridge National Laboratory, Oak Ridge, Tennessee 37830 (207)

GARY D. STONER,[1] Human Tissue Studies Section, Laboratory of Experimental Pathology, National Cancer Institute, National Institutes of Health, Bethesda, Maryland 20205 (15, 65)

GARY E. STRIKER, Departments of Pathology and Medicine, University of Washington, Seattle, Washington 98195 (135)

IRA J. STRUMPF, Pulmonary Branch and Pathology Branch, National Heart, Lung, and Blood Institute, National Institutes of Health, Bethesda, Maryland 20205 (95)

SUSAN V. SZAPIEL, Pulmonary Branch and Pathology Branch, National Heart, Lung, and Blood Institute, National Institutes of Health, Bethesda, Maryland 20205 (95)

BENJAMIN F. TRUMP, Department of Pathology, University of Maryland, Maryland Cancer Program, and Maryland Institute for Emergency Medicine, Baltimore, Maryland 21201 (1)

AURORA VIAJE, Biology Division, Oak Ridge National Laboratory, Oak Ridge, Tennessee 37830 (207)

[1]*Present address:* Department of Pathology, Medical College of Ohio, Toledo, Ohio 43619.

KERN WILDENTHAL, Pauline and Adolph Weinberger Laboratory for Cardiopulmonary Research, Departments of Physiology and Internal Medicine, University of Texas Health Science Center at Dallas, Dallas, Texas 75235 (167)

STUART H. YUSPA, *In Vitro* Pathogenesis Section, Laboratory of Experimental Pathology, National Cancer Institute, Bethesda, Maryland 20205 (289)

PREFACE

One of the central problems in biomedical research is the extrapolation of data from experimental animals to the human situation. This is especially a problem in studies of chronic diseases such as atherosclerosis, diabetes, and cancer. Extrapolation is made difficult by variation among species as well as variation among individuals within a species. Such variation may be wide, particularly in outbred populations such as humans. Finally, within a single individual one may find variations in response to exogenous agents among different tissues and within a tissue among different cell types.

The use of cultured human tissues and cells for investigations in biomedical research can be a logical bridge between experimental animals and the beneficiary of our research, humans. Volumes 21A and 21B of *Methods in Cell Biology* describe the current state of methodology in culturing human tissues and cells. The organization is on the basis of organ systems and both explant and cell culture approaches are included. A common strategy in this field of research is initially to develop the methodology with cells and tissues from experimental animals and then to adapt this methodology for use with human cells and tissues. Thus far, some investigations have been conducted only in cells and tissues from experimental animals. Studies with human material are still at an early stage of development. A few chapters are included that reflect the progression in methodology from experimental animals to humans. In general, this progression has been easier than most of us would have predicted.

The primary objective of these methodologic studies is to provide experimental tools for investigating the pathogenesis of human disease. In fact, as noted in many of the chapters, cultured human cells and tissues are already being used in conjunction with sophisticated methods to investigate a wide variety of important problems in biomedical research. The purpose of these volumes is to further encourage future investigations.

CURTIS C. HARRIS
BENJAMIN F. TRUMP
GARY D. STONER

Chapter 1

Methods of Organ Culture for Human Bronchus[1]

BENJAMIN F. TRUMP, JAMES RESAU,
AND LUCY A. BARRETT

Department of Pathology,
University of Maryland,
Maryland Cancer Program, and
Maryland Institute for Emergency Medicine,
Baltimore, Maryland

I. Introduction

One goal of organ culture is the maintenance of *in vivo* characteristics and specific cellular relationships. Cellular proliferation is not encouraged other than

[1]Supported in part by Contract NO1-CP-43237 from the National Cancer Institute. This is Contribution 786 from the Cellular Pathobiology Laboratory, University of Maryland School of Medicine, Baltimore, Maryland.

1

for normal maintenance of the tissue but may be useful for specific purposes, e.g., carcinogenesis studies. The epithelial component of the explant initiates wound repair when the tissue is cut into pieces as it is put into culture. This phenomenon of wound repair probably has more to do with changing the morphology and direction of differentiation within the explant than any other factor or condition of culture.

The first tissue to be maintained in explant organ culture, consisting of embryonic limb buds, was reported over 50 years ago (Strangeways and Fell, 1926). A major advance in the *in vitro* maintenance of adult tissue came with the demonstration that mature rat tissue could survive over 1 week in culture (Trowell, 1959).

Methods for the long-term explant culture of human bronchi have been developed recently by Trump *et al.* (1974) and Barrett *et al.* (1976), using techniques adapted from those described by Clamon *et al.* (1974) for hamster trachea. Although several investigators in the past have successfully cultured portions of human bronchi, they did so only for short periods of a few days to a few weeks (Lasnitzki, 1956; Cailleau *et al.*, 1959; O'Donnell *et al.*, 1973). At the present time we are able to maintain bronchial explants in culture for as long as 1 year, and it seems likely from our present data that even longer periods could be obtained if desired. Furthermore, it has been possible to xenotransplant cultures of human bronchi successfully into nude mice at the time of obtainment or after various periods of *in vitro* culture (Valerio *et al.*, 1980).

II. Methods

A. Obtainment

One of the fundamental prerequisites for maintaining any human tissue *in vitro* is to obtain the tissue in a viable state. We have developed two means for achieving this goal. The first source is tissue removed at the time of surgical resection for neoplastic or nonneoplastic disease, and the second is tissue obtained at the time of immediate autopsy. The latter is especially important for studies on the mechanisms of carcinogenesis, as this patient population consists predominantly of young adult individuals suffering from head trauma who are therefore predominantly free of chronic disease.

The method of obtaining surgical tissues involves deploying collection teams to the operating room at the time of surgery. The tissues are dissected as soon as possible, under the supervision of diagnostic anatomic pathologists, usually within 30 minutes after resection. The bronchi are freed of adhering lung parenchyma and washed in sterile L-15 medium. The tissues are then placed in containers containing chilled L-15 medium, which are maintained at 0–4°C for transport to the tissue and cell culture laboratory. Experiments in our laboratory have shown that human bronchi can survive total ischemia at 37°C for up to approximately 3 hours, while at room temperature they can survive for several days and for 1 week at 0–4°C (Barrett *et al.*, 1977). This makes possible the development of numerous collaborative studies involving the transport of human tissue for rather long distances.

The immediate autopsy is a procedure developed at our institution for rapid obtainment of tissues prior to ischemic change. These autopsies are performed within 5–30 minutes after death, predominantly on individuals who have succumbed to head injury. Various on-call personnel, trained in autopsy procedure and tissue culture, are deployed to the site of the autopsy at moments' notice, day or night. The autopsy is performed under the auspices of the Medical Examiner of the State of Maryland. The details of this procedure have been described previously (Trump *et al.*, 1973), and the methods for handling the tissue are the same as those described above.

B.   Culture Methods

1.   GENERAL

Both the human and bovine trachea and bronchus were cultured as 1-cm-square pieces in 60-mm Falcon plastic petri dishes with the epithelial surface uppermost. The culture medium was added to just reach, but not cover, the epithelial surface of the explant. The dishes were incubated at 36.5°C, and the medium was changed two times per week.

2.   ENRICHED MEDIUM

The nutritionally complex medium CMRL-1066 (Gibco) was supplemented with 0.1 μg/ml hydrocortisone, 1.0 μg/ml insulin, 2 mM L-glutamine, 5% heat-inactivated fetal calf serum, 200 U/ml penicillin, 200 μg/ml streptomycin, and 2.5 μg/ml amphotericin B. The human explants cultured in this medium were placed in controlled-atmosphere chambers (Bellco Company, Inc., Vineland, New Jersey) which were gassed with a mixture of 45% oxygen, 50% nitrogen, and 5% carbon dioxide and placed on rocker platforms which rocked at 3–4 cpm as previously described (Barrett *et al.*, 1976).

3.   BASAL MEDIUM

The basal medium, modified Eagle's medium (MEM) with Hanks' salts, was supplemented with 1.0 μg/ml insulin, 2 mM L-glutamine, 10% heat-inactivated fetal calf serum, 200 U/ml penicillin, 200 μg/ml streptomycin, and 2.5 μg/ml amphotericin B. The dishes were incubated in an atmosphere of 5% carbon dioxide and air with 100% humidity.

C. Xenografts

Bronchial explants were grafted into nude mice as discussed elsewhere in this volume. This was associated with the preservation and formation of well-differentiated epithelium (often surrounding a cystic space) after an initial period of inflammation and vascularization. Maintenance for over 400 days has been achieved (Valerio *et al.*, 1980).

III. Results

A. Normal Human Bronchial Epithelium

Normal human bronchial epithelium is composed of a single layer of pseudo-stratified cells. The tall, columnar ciliated cells are the most numerous. Tall,

Fig. 1. Human bronchus fixed after 2 months in culture. Several ciliated cells are seen among nonciliated cells with long microvilli. Weblike strands of mucus can be seen at the lower left. SEM. Bar represents 1 μm. $\times 5000$.

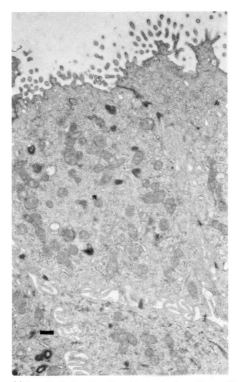

Fɪɢ. 2. Human bronchial epithelial cells after 11 weeks in culture. The surface of the cells contains numerous long microvilli. Mucous droplets are aligned at the luminal aspect of the cells. The cell shape has changed from tall columnar to cuboidal. TEM. Bar represents 2 μm. ×4000.

columnar mucous cells are of two types: "goblet" cells whose apices are distended with large mucous granules and "small mucous granule cells" (SMGCs) which have tiny mucous granules in the apex which are invisible by light microscopy unless stains for mucosubstances are used. Below the columnar cells the basal cells, not numerous in normal epithelium, rest on a thin basal lamina between the bases of the mucous and ciliated cells which also rest on the basal lamina. Cytologically, the basal cells are quite similar to the mucous cells. Least frequent are the "endocrine" cells (K cells, neurosecretory cells). These cells also rest on the basal lamina and sometimes reach the lumen and are characterized by numerous dense-core granules.

Of these components, the ciliated cells are apparently end-stage and cannot divide. The other cellular constituents can divide and do so when appropriately stimulated. In some instances, mucous-containing cells develop cilia.

When tracheobronchial mucosa was injured, as for example, by a blunt probe as described by McDowell *et al.* (1979), the ciliated cells were lost, whereas the mucous and basal cells flattened to cover partially the denuded basal lamina. Subsequently, the defect was covered by one or two layers of simple squamous cells. The cells covering the defect divided. Eventually the defect was covered by four or five layers of cells, some of which were typical or nonkeratinizing epidermoid metaplasia. The presence of undifferentiated (indifferent) cells at 72 hours postinjury eventually gave way to a fully differentiated epithelium at 96 hours.

B.   Human Bronchus in Enriched Medium

Differentiated ciliated columnar cells could be observed up to 2 months in culture (Fig. 1), although there was an overall tendency for a reduction in the height of the columnar cells with time in culture (Fig. 2). Goblet cells were very rare, although SMGCs were common. After 4 months of culture the epithelium consisted of two to three layers of cells which formed a nonkeratinizing partially

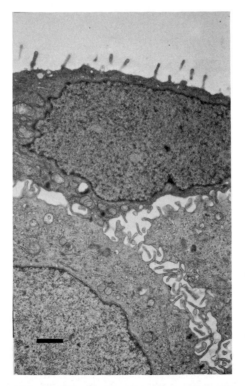

Fig. 3.   Human bronchial epithelial cells after 4 months in culture. The surface cells are flat. Long microvilli persist on the cell surface. TEM. Bar represents 1 μm. ×7000.

Fig. 4. Human bronchial epithelium after 1 year in culture. The surface of the explant is covered with flat epidermoid cells containing various densities of microvilli. SEM. Bar represents 1 μm. ×3000.

squamous epithelium (Fig. 3). Many cells contained small mucous granules, and the surface was covered with short microvilli.

The surface morphology remained relatively unchanged for up to 1 year of culture. A one- to two-cell layer covered the explant proper, which at this time showed extensive subepithelial necrosis and was essentially acellular except for a few persistent submucosal glands. The epithelium of the submucosal glands was low cuboidal and continued to manufacture an acidic mucous substance identified by a positive periodic acid Schiff–alcian blue (pH 2.5) reaction.

By scanning electron microscopy (SEM), the surface of the explants was seen to consist of large, flat cells covered with microvilli of various sizes and densities (Fig. 4). The morphology was consistent with squamous or epidermoid metaplasia.

C.   Human Bronchus in Basal Medium

Human bronchus maintained up to 5 months in MEM consisted of roughly equal numbers of ciliated and nonciliated cells. All the nonciliated cells were covered with microvilli, and some contained enough mucus to cause the luminal aspect of the cell to bulge slightly (Fig. 5). The epithelium in the center of the explant varied from columnar to cuboidal in shape. At the cut edges of the explant the frequency of ciliated cells was diminished as migrating and proliferating basal and SMGCs attempted to repair the "wound." Conversion of proliferating epidermoid cells to columnar cells was not observed even after 5 months. This may have been due to the fact that at the edge of the explant the wound is "infinite" in size, so that proliferation and migration never cease in this *in vitro* system.

D.   Bovine Bronchus in Basal Medium

Beating cilia were observed on the edges of the explants continuously throughout the year-long culture. Initially and up to about 4 months, the cilia were uniform in density and were synchronously beating. Samples fixed at this time and examined by SEM were seen to be covered by a ciliated epithelium.

By 5 months of culture areas lacking cilia were observed at the edge of the

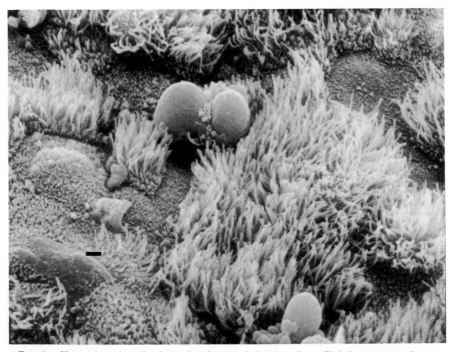

FIG. 5.   Human bronchus after 5 months of culture in basal medium. Globular mucus can be seen protruding from the cells. SEM. Bar represents 1 μm. ×4000.

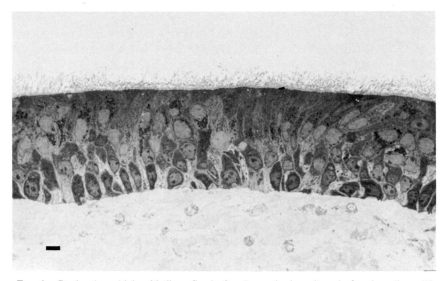

Fig. 6.   Bovine bronchial epithelium fixed after 5 months in culture in basal medium. The luminal cells have remained ciliated and columnar. Stained with hematoxylin and eosin. Light microscopy. Bar represents 4 μm. ×1000.

explant using the inverted microscope. By SEM the areas lacking cilia were seen to be composed of elongated nonciliated cells, apparently representing proliferating epithelial cells. Light microscopy of the center of the explant revealed tall, ciliated columnar epithelium (Fig. 6).

Clumps or aggregates of beating cilia were observed on the edge of the explants using phase-contrast optics up to the 365th day of culture when the last sample was fixed. In the center of the explant, the tall, ciliated columnar epithelium was retained for a 12-month culture period (Figs. 7–9). The cells on the edges of the explant had more sparse cilia (Fig. 10), and the surface of many cells had microvilli of various lengths.

IV.   Discussion

These findings clearly demonstrate the feasibility of long-term explant culture of human bronchial epithelium. Although all the critical factors are by no means understood, it is evident that variation in cellular differentiation is expressed. In explants in the CMRL-1066 medium rocked in an atmosphere of 45% oxygen as described above, the epithelium remained columnar for only about 1–2 months. After this time it became squamous and continued to proliferate. In the explants

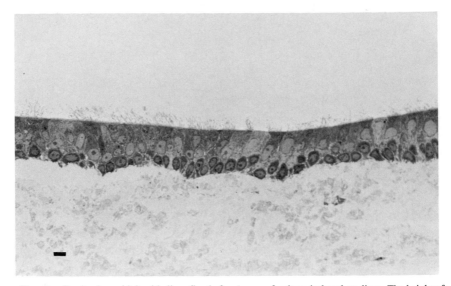

FIG. 7. Bovine bronchial epithelium fixed after 1 year of culture in basal medium. The height of the luminal cells has diminished somewhat but remains columnar. The interior of the explant proper is completely acellular. Stained with hematoxylin and eosin. Light microscopy. Bar represents 4 μm. ×750.

FIG. 8. Bovine epithelium fixed after 1 year in basal medium. The surface of the center of the explant is covered by numerous cilia which obscure the surface of the explant. SEM. Bar represents 4 μm. ×1000.

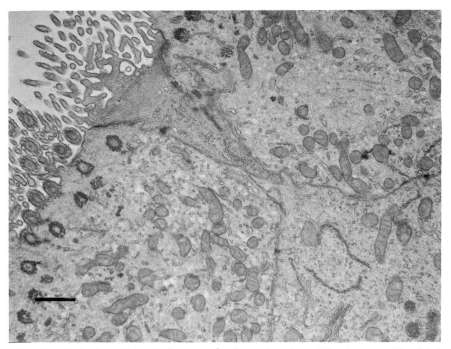

FIG. 9. Bovine bronchial epithelial cells after 1 year in culture. Ciliated and nonciliated (microvillous) cells can be distinguished. TEM. Bar represents 1 μm. ×10,000.

maintained in the simpler MEM medium, normal differentiation was maintained for up to 1 year.

It has been established in other studies from our laboratory that squamous metaplasia is part of the normal repair reaction of the bronchial epithelium to simple mechanical injury. Such metaplasia also occurs following treatment of the bronchial epithelium with carcinogens, although the total relationship between the two cases is not clear. In the culture situation, therefore, it is possible that the culture condition employing the enriched CMRL medium is in some way injurious or otherwise stimualtes continual "repair-type" responses. The xeno-transplant studies clearly show that the culture-induced metaplastic changes are reversible, because when such CMRL-cultured explants are transplanted into nude mice, the epithelium reverts to normal and columnar differentiation takes place and is maintained for over 1 year.

In somewhat similar experiments, Mossman et al. (1978), studying hamster tracheal explants, has also noted that normal epithelium undergoes squamous metaplasia in organ cultures maintained in a complex, chemically defined medium such as Waymouth MAB87/3. In comparing these results to those with MEM which allowed the maintenance of normal differentiation, similar to that

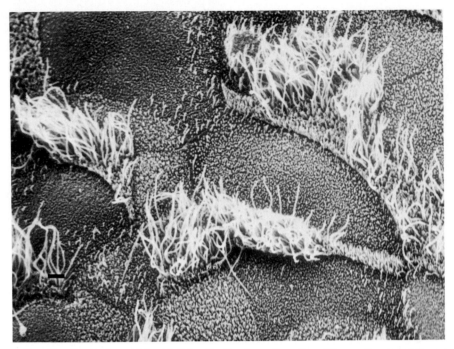

Fig. 10. Bovine bronchial epithelial cells after 1 year in culture. These cells covered the cut edges of the explant. SEM. Bar represents 2 μm. ×2000.

described by us, these investigators attributed the more rapid growth and metaplasia to the presence of several nonessential amino acids of which L-Glu and L-Ser were deemed most influential in inducing metaplastic changes in separate tests. It furthermore suggested that the presence of these nonessential amino acids stimulated DNA and protein synthesis, increasing cell division and producing a relative vitamin A deficiency, since the hormone was not added to the medium. Our results, however, suggest that vitamin A may not be the primary cause. Further experiments will clearly be necessary to clarify these problems. Other factors might induce differences in oxygen levels, agitation, amount of hydrocortisone, etc. It is furthermore important to establish whether or not the enriched medium that stimulates growth is best suited for studies on carcinogenesis.

V.   Perspectives

When cells were first grown *in vitro* in 1913 in the laboratory of Alex Carrell, he concluded that the life span of the cultured cells was infinite (Carrel, 1913).

Subsequently, Hayflick (1965) reported that diploid cells had a finite lifetime *in vitro* which may be an expression of aging or senescence at the cellular level. Bell *et al.* (1978) suggest that diploid cells in culture tend toward differentiation which eliminates the opportunity to divide. However, given the right conditions, the intrinsic capacity to remain immortal might be a property of a limited sub-population of normal cells in culture. Obviously, the question of the life span of cells *in vitro* remains to be answered.

With the advent of improved media and techniques for the maintenance of large portions of organs for extended periods of time, the questions concerning the life span of cells can now be asked of tissues in organ culture. Such studies become more complex because not only are explants composed of different cell types but cells of similar types are experiencing a different environment at any one moment. For example, the epithelial cells at the cut edge are somehow stimulated to begin wound repair. These cells begin to flatten and eventually enter the mitotic phase of the cell cycle. Other epithelial cells at the center of the explant remain quiescent and differentiated. Apparently here cell division occurs only in the event of normal cell loss due to aging or local injury.

The use of organ culture to study some of the major questions of cell biology has only recently been initiated. The amount of information available to be understood and deciphered awaits only the imagination and attention of cell biologists. Many questions concerning carcinogenesis, aging, differentiation, tissue interactions, and morphogenesis can now begin to be answered with the help of organ culture methodology.

REFERENCES

Barrett, L. A., McDowell, E. M., Frank, A. L., Harris, C. C., and Trump, B. F. (1976). *Cancer Res.* **36**, 1003–1010.

Barrett, L. A., McDowell, E. M., Harris, C. C., and Trump, B. F. (1977). *Beitr. Pathol.* **161**, 109–121.

Bell, E., Marek, L. F., Levin, D. S., Merrill, C., Sher, S., Young, I. T., and Eden, M. (1978). *Science* **202**, 1158–1163.

Cailleau, R., Crocker, T. T., and Wood, D. A. (1959). *J. Natl. Cancer Inst.* **22**, 1027–1037.

Carrel, A. (1913). *J. Exp. Med.* **18**, 287–292.

Clamon, G. H., Sporn, M. B., Smith, J. M., and Saffiotti, U. (1974). *Nature (London)* **250**, 64–66.

Hayflick, L. (1965). *Exp. Cell Res.* **36**, 614–636.

Lasnitzki, I. (1956). *Br. J. Cancer* **10**, 510–516.

McDowell, E. M., Becci, P. J., Schürch, W., and Trump, B. F. (1979). *J. Natl. Cancer Inst.* **62**, 995–1008.

Mossman, B. T., Heintz, N., MacPherson, B. U., and Craighead, J. E. (1978). *Proc. Soc. Exptl. Biol. Med.* **157**, 500–505.

O'Donnell, T. V., Crocker, T. T., and Nunes, L. L. (1973). *Cancer Res.* **33**, 78–87.

Strangeways, T. S. P., and Fell, H. B. (1926). *Proc. R. Soc. London* **99**, 340–366.

Trowell, O. A. (1959). *Exp. Cell Res.* **16**, 118–147.

Trump, B. F., Valigorsky, J. M., Dees, J. H., Mergner, W. J., Kim, K. M., Jones, R. T., Pendergrass, R. E., Garbus, J., and Cowley, R. A. (1973). *Hum. Pathol.* **4,** 89–109.

Trump, B. F., McDowell, E. M., Barrett, L. A., Frank, A. L., and Harris, C. C. (1974). "Experimental Lung Cancer. Carcinogenesis and Bioassays." Springer-Verlag, Berlin and New York.

Valerio, M. G., Fineman, E. L., Bowman, R. L., Harris, C. C., Trump, B. F., Hillman, E. A., and Heatfield, B. M. (1980). *Proc. Int. Workshop Nude Mice, 3rd, 1979,* Gustav Fischer Verlag (in press).

Chapter 2

Identification and Culture of Human Bronchial Epithelial Cells

GARY D. STONER,[1,3] YOICHI KATOH,[1]
JEAN-MICHEL FOIDART,[2] GWENDOLYN A. MYERS,[1]
AND CURTIS C. HARRIS[1]

I. Introduction

The development of methods for culturing human bronchial epithelial cells would permit analysis of factors that influence their normal differentiation and of the effects of chemical carcinogens and other toxins on these important target cells in human lung disease. During the last two decades, there have been several

[1]Human Tissue Studies Section, Laboratory of Experimental Pathology, National Cancer Institute, National Institutes of Health, Bethesda, Maryland 20205.

[2]Laboratory of Developmental Biology and Anomalies, National Institute of Dental Research, National Institutes of Health, Bethesda, Maryland 20205.

[3]Present address: Department of Pathology, Medical College of Ohio, 3000 Arlington Ave., Toledo, Ohio 43614.

2. Cytochemistry

Special stains used to identify products of the cells in primary culture were alcian blue–periodic acid Schiff (PAS) for acidic and neutral mucopolysaccharides (mucus) and glycogen, orange G–alcian blue for prekeratin and keratin, oil red O for lipid, and a silver impregnation stain for reticulin (Berman *et al.*, 1978). Ascorbic acid (50 μg/ml) was added to the cultures daily for a period of 7 days prior to staining with the silver impregnation stain.

3. Immunological Markers

Immunological methods for the detection of blood group antigens, collagen type, fibronectin, and keratin are depicted in Table I and have been described in detail by Katoh *et al.* (1979). We used indirect immunofluorescence (IF) methods for the detection of blood group antigens, collagen type, and fibronectin, and the triple-layer peroxidase–antiperoxidase (PAP) method (Sternberger *et al.*, 1970) for keratin, however, either method can be used to detect these antigens. Procedures used were as follows.

  a.   *Blood Group Antigens.*   Bronchial explants (0.2 × 0.2 cm) were cul-

TABLE I

Immunofluorescence and Immunoperoxidase Staining of Cell Antigens

Antigen	Method[a]	Primary antibody[b]	Secondary antibody[b]	Tertiary antibody[b]
Blood groups A, B, and H	IF	Human anti-A, -B, and -H sera (1:1)[b]	Goat anti-human IgG (1:10)[c]	
Collagens, types I, II, III, and IV	IF	Rabbit anticollagen sera (protein conc. 100 μg/ml)	Goat anti-rabbit IgG (1:30)[c]	
Fibronectin	IF	Rabbit antifibronectin serum (protein conc. 100 μg/ml)	Goat anti-rabbit IgG (1:30)	
Keratin	PAP	Rabbit anti-human keratin serum (1:50)	Goat anti-rabbit IgG (1:20)	Rabbit PAP (1:50)

  [a] IF, Indirect immunofluorescence; PAP, triple-layer PAP method.
  [b] Numbers in parentheses indicate dilution of antisera.
  [c] Conjugated with fluorescein isothiocyanate.

tured on glass coverslips in 60-mm plastic dishes containing 3–4 ml of culture medium. After 2–3 weeks in culture, when the diameter of the outgrowth from the explant periphery was approximately 1–3 cm, the coverslips were rinsed with two changes of PBS and fixed with ice-cold acetone for 5 minutes. After the coverslips were dried in air, they were immersed in PBS and then removed; the cellular outgrowth was then stained for A, B, and H blood group antigen-specific reactivity. The coverslips with attached cells were incubated with various dilutions of the primary antibodies, followed by exposure to the secondary antibodies conjugated with fluorescein isothiocyanate. The reaction was examined with a Leitz Ortholux-2 fluorescence microscope. Secondary and tertiary subcultures of bronchial cells in plastic flasks were fixed in ice-cold ethanol and also stained by indirect immunofluorescence for blood group antigens.

b. Collagen. Outgrowths of bronchial cells on coverslips were obtained as described above. The outgrowths were fixed in ice-cold acetone, washed in PBS, dried, and reacted by indirect immunofluorescence with antibodies to types I, II, III, and IV procollagen and collagen. The procollagens and collagens were isolated and purified as described by Foidart *et al.* (1978). Antibodies prepared in rabbits were purified by cross-immunoabsorption (Nowack *et al.*, 1976), and monospecificity was verified by immunofluorescence radioimmunoassay and hemagglutination assays (Foidart *et al.*, 1978; Yaoita *et al.*, 1978).

c. Fibronectin. Outgrowths of bronchial cells, fixed and washed as described above, were reacted by indirect immunofluorescence with rabbit antibodies against purified fibronectin from fresh human serum. The fibronectin was isolated by affinity chromatography (Engvall and Rouslahti, 1977). The protein eluted with 1 M potassium bromide was electrophoresed on 5% acrylamide gels, and the 220,000-dalton monomer was eluted from crushed acrylamide and used as a source of antigen. Antibodies were prepared in rabbits, and monospecificity verified by radioimmunoassay as described by Foidart *et al.* (1979).

d. Keratin. Outgrowths of bronchial cells on coverslips were fixed in cold acetone and washed in PBS as described above. In addition, exfoliated cells released into the medium during culture were centrifuged onto slides with a cytocentrifuge, fixed in cold acetone, washed, and dried. The coverslips and slides were stained by the PAP method using antibodies raised in rabbits against prekeratin proteins isolated from stratum corneum from human leg or fibrous protein from human hair (Lee *et al.*, 1976). (These antibodies were kindly supplied by Howard Baden, Department of Dermatology, Harvard Medical School.) To detect keratin, the specimens were reacted with the primary rabbit antibody, then with goat anti-rabbit IgG, and finally with rabbit PAP. Then a freshly prepared solution of 3,3′-diaminobenzidine tetrachloride (DAB, 0.6 mg/ml) containing 1 μl/ml of 30% hydrogen peroxide in PBS (pH 7.6) was added to the specimen for 8 minutes. The keratin was localized by a brown staining reaction. The reaction is due to oxidation of the DAB by the enzyme horseradish peroxidase.

C.   Subcultures

Preliminary experiments were conducted in attempts to establish conditions for the routine subculture of bronchial epithelial cells. During the early stages of primary culture, epithelial cells and fibroblasts remain physically separated; therefore, contaminating fibroblasts can be scraped from the surface of the dish with a rubber policeman and relatively purified populations of epithelial cells made available for transfer. The following conditions were examined for subculture of bronchial epithelial cells: (1) Cells were subcultured after different periods in primary culture. (2) Various enzyme solutions (0.05% trypsin–EDTA, 0.2% pronase) and chelating agents (0.02% EGTA, 0.02% EDTA) were used for cell dispersion. In addition, the cells were scraped into the culture medium with a rubber policeman and dispersed with a pipet to a suspension of small aggregates and single cells before transfer. (3) Cells were subcultured into standard plastic dishes, or into dishes coated with either type I or type IV collagen (Murray *et al.*, 1979) or containing various numbers of irradiated (2000 R) IMR-90 human lung fibroblasts (Nichols *et al.*, 1977), i.e., 5×10^4, 1×10^5, or 5×10^5 cells per dish. (4) After subculture, cells were fed with either fresh medium or "conditioned" medium, i.e., medium exposed to primary explant cultures for 36 hours and then mixed 1:1 with fresh medium.

III.   Results

A.   Morphology and Growth of Bronchial Cells in Primary Culture

1.   Light Microscopy

Primary cultures of bronchial cells were established from tissues taken from 28 of 32 patients. When explants of human bronchus were placed in the center of tissue culture dishes, polygon-shaped epithelial cells (Fig. 1A) grew out from the explant periphery onto the culture dish before the fusiform-shaped fibroblast cells (Fig. 1B), and on the side of the explant opposite the fibroblasts. Usually, there was substantial outgrowth of both epithelial and fibroblast cells after 2 weeks in culture. Considerable variability was observed in the extent of epithelial outgrowth from tissues taken from different patients. In most cases, epithelial cells

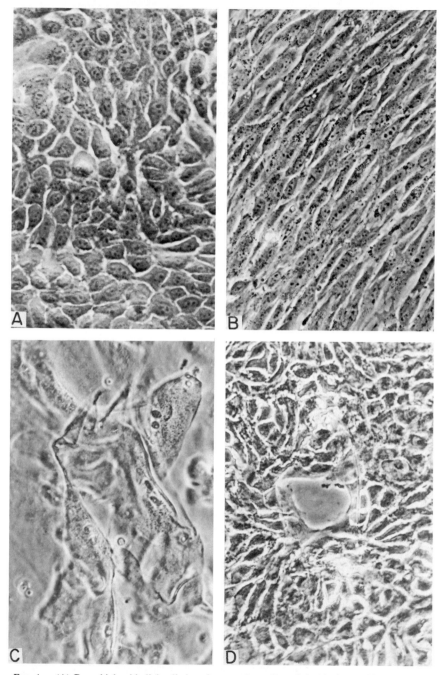

Fig. 1. (A) Bronchial epithelial cells in primary culture; 2-week incubation. ×220. (B) Bronchial fibroblasts in primary culture; 2-week incubation. ×220. (C) Squamous epithelial cells sloughed into the culture medium. ×440. (D) Hemicyst in epithelial area of 30-day primary culture. ×200.

grew to a distance of 1–2 cm from the explant, and then lateral outgrowth either stopped or was inhibited by physical contact of the epithelial cells with fibroblasts. However, in other cases, there was either no outgrowth or growth that covered the entire surface of the dish.

Observation of plastic-embedded specimens indicated that both epithelial and fibroblast cells could stratify to densities of several cell layers in thickness (Stoner *et al.*, in press). In the epithelial outgrowth, [^{3}H]thymidine was incorporated predominately into the nuclei of cells attached to the plastic dish and to a lesser extent into the nuclei of cells in the upper layers of the multilayer. Cells in all layers of the fibroblastic outgrowth appeared to be equally capable of incorporating [^{3}H]thymidine into their nuclei. The average labeling index of fibroblast and epithelial cells has been examined in three cultures maintained for 25 days. The labeling index of fibroblast cells [14.0 ± 6.7% (mean ± S.D.)] exceeded that of epithelial cells (8.2 ± 4.2%). Epithelial cells at a distance from the explant had a higher labeling index (9.7 ± 3.8%) than epithelial cells in close proximity to the explant (5.6 ± 3.6%). The average labeling index in epithelial cells of the explants from these three cultures was not determined but, in other experiments, the labeling index of epithelial cells in the explants was 2.9 ± 1.6%.

Ciliary activity, detected by localized movement of culture medium over the epithelial cells in the explant and in the outgrowth, was observed in cultures maintained for at least 6 months. Several conclusions could be made with regard to ciliary movement in the epithelial cells: (1) It was dependent upon the presence of the explant in the culture; usually within 24 hours after removal of the explant, ciliary beating ceased. (2) It was frequently observed in the cells of the explant and not in the epithelial outgrowth from the same explant. (3) There was a marked variation in the proportion of epithelial cells in the outgrowth that exhibited ciliary movement. (4) The beating of the cilia was more vigorous in cells in close proximity to the explant than in cells at a distance from the explant. (5) It decreased with increasing age of the culture. (6) It was observed in a variable proportion (0–100%) of epithelial outgrowths from different explants collected from the same patient.

With increasing age of the culture, the majority of epithelial cells in the outgrowth differentiated progressively to a squamous keratinizing epithelium. Squamous epithelial cells were observed on the surface of the cultures, and they continuously sloughed into the medium (Fig. 1C). Hemicysts were observed in older cultures and were formed by the detachment of squamous cells from the basal cell layers (Fig. 1D).

As fibroblast cells grew out from the explants, they came into contact with the epithelial cells and prevented further outgrowth of the epithelium. It was possible to reinitiate epithelial outgrowth by simply scraping the fibroblasts from the surface of the dish with a rubber policeman. Epithelial cells grew readily in areas of the dish that had been occupied by fibroblasts.

As indicated in Table II, cytochemical stains were useful for identifying bronchial epithelial cells and fibroblasts in primary culture. Epithelial cells, and not fibroblasts, stained positively with alcian blue–PAS before and after treatment with diastase, indicating the production of acidic and neutral mucopolysaccharides, two components of mucus. There was no histochemical evidence for the production of glycogen in either epithelial or fibroblast cells. Squamous epithelial cells on the surface of the multilayer and exfoliated cells stained positively (orange to magenta) with the orange G–alcian blue stain for prekeratin and keratin; other epithelial cells and all fibroblasts were negative. In addition, when the cultures were grown in medium containing ascorbic acid, only fibroblast cells produced reticulin fibers stained by the silver impregnation method (Berman et al., 1978). These fibers had the appearance of collagen in the electron microscope (Berman et al., 1978).

2.   ELECTRON MICROSCOPY

The ultrastructural morphology of epithelial cells and fibroblasts maintained for 2 and 4 weeks in primary culture was determined. Epithelial cells in 2-week cultures were usually stratified to densities of two to three cell layers in thickness (Fig. 2). The cells in the bottom layer, adjacent to the dish, were more rounded and usually more electron-dense than those in the upper layers. These cells contained the Golgi complex, microvilli, mitochondria, tonofilaments, 100-Å filaments, and rough endoplasmic reticulum and were joined by desmosomal junctions. Membrane-bound inclusions resembling mucus droplets were observed in a small proportion of these cells. Cells in the upper layers of 2-week cultures were of two morphological types: (1) large, flattened cells with an ovoid nucleus, and (2) smaller, electron-lucent, flattened cells. The large cells contained cytoplasmic organelles, but most of the smaller, electron-lucent cells had

TABLE II

PRODUCTS OF BRONCHIAL CELLS IN CULTURE DETECTED BY CYTOCHEMICAL METHODS

Stain	Product	Cell type	
		Epithelial	Fibroblast
Alcian blue–PAS	Acidic and neutral mucopolysaccharides	+	−
Orange G–alcian blue	Prekeratin and keratin	+	−
Silver impregnation	Reticulin (collagen)	−	+

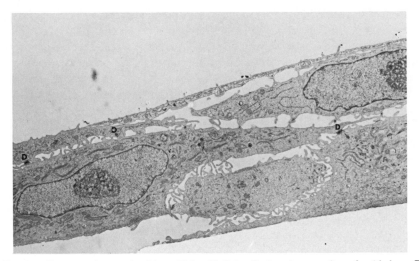

Fig. 2. Electron micrograph of bronchial epithelial cells in primary culture for 14 days. D, Desmosome. ×9000.

no distinguishable cytoplasmic organelles. Both morphological types were joined together by junctional complexes including tight junctions and desmosomes, and to the underlying cells by desmosomes.

In 4-week cultures, the squamoid features of the epithelial outgrowth were even more apparent than in 2-week cultures (Fig. 3). There were frequently one to two layers of electron-lucent squamous cells on the medium surface of the multilayer, which contained few, if any, cytoplasmic organelles. Cells in the middle layer contained abundant tonofilaments which either encircled the nucleus or were randomly distributed throughout the cytoplasm. The majority of the cells in the basal layer, adjacent to the plastic dish, contained either only a few tonofilaments or no tonofilaments. Some of the basal cells had an extensive ribosomal network and contained membrane-bound mucus-like granules suggestive of secretory activity. Basal lamina and hemidesmosomes were not observed in the epithelial outgrowth. Cilia were seen in a small proportion of the cells examined.

Unlike the epithelial cells, bronchial fibroblasts were not joined by junctional complexes, including desmosomes. Lipid droplets were frequently observed in older primary cultures of fibroblasts.

3. Immunological Markers

Epithelial cells initiated from explants from blood type-A or -B patients possessed A or B blood group antigen-specific reactivity for periods of at least 90 days in primary culture (Fig. 4A). Fibroblast cells were negative (Fig. 4B). In addition, epithelial cells could be distinguished from fibroblasts in secondary and tertiary cultures of mixed cell types, since only the epithelial cells reacted with antibodies to blood group A or B antigens. Under our culture conditions, epithelial cells from blood group H (type O) patients did not react with "anti-H" serum obtained from a patient with Bombay blood group Oh or "anti-H" lectin from *Ulex europaeus* (Boyd and Shapleigh, 1954).

As expected, only the fibroblast cells reacted with antibodies to types I and III collagen, and epithelial cells were negative (Fig. 4C and D). Epithelial cells reacted weakly with antibody to type IV collagen (Fig. 4E), and fibroblasts were negative (Fig. 4F). In some cultures, a small number of cells close to the area of the dish where the explant was located reacted with antibody to type II procollagen, suggesting that these cells were chondrocytes derived from the cartilage in the explant (Miller and Matukas, 1974).

Squamous cells on the medium surface of the epithelial outgrowth were stained by the PAP method, using rabbit antibody to human skin prekeratins (Fig. 5A), and not by normal rabbit serum (Fig. 5B). Fibroblast cells were not stained by either serum.

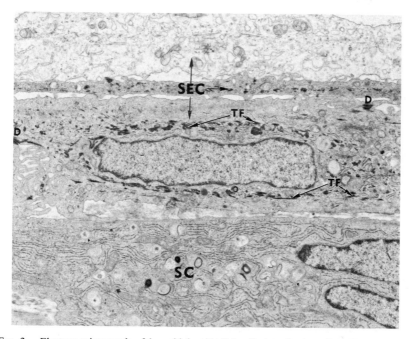

FIG. 3. Electron micrograph of bronchial epithelial cells in primary culture for 25 days. D, Desmosome; TF, tonofilaments; SC, secretory cell; SEC, three squamous epithelial cells. ×7000.

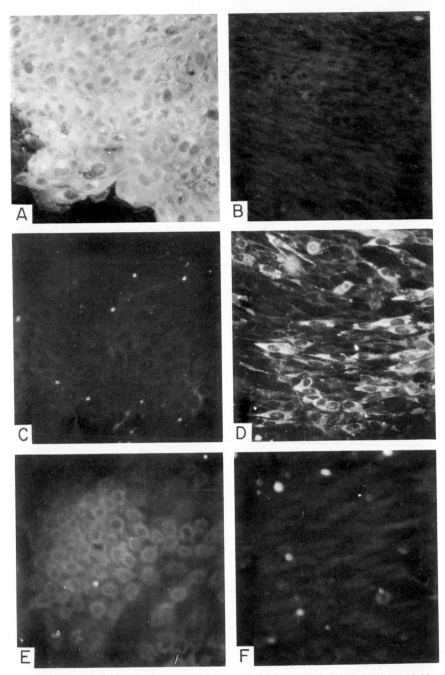

FIG. 4.   (A) Positive immunofluorescence in bronchial epithelial cells from a patient with blood type A after reaction with anti-A serum. (B) Lack of fluorescence in fibroblasts from same patient after treatment with anti-A serum. (C) Epithelial cells do not react with antibodies to type I collagen. (D) Fibroblasts react with antibodies to type I collagen. Fibroblasts were also positive for type III collagen, and epithelial cells were negative. (E) Epithelial cells react weakly with antibodies to type IV collagen. (F) Fibroblasts do not react with antibodies to type IV collagen. (A–F) ×220.

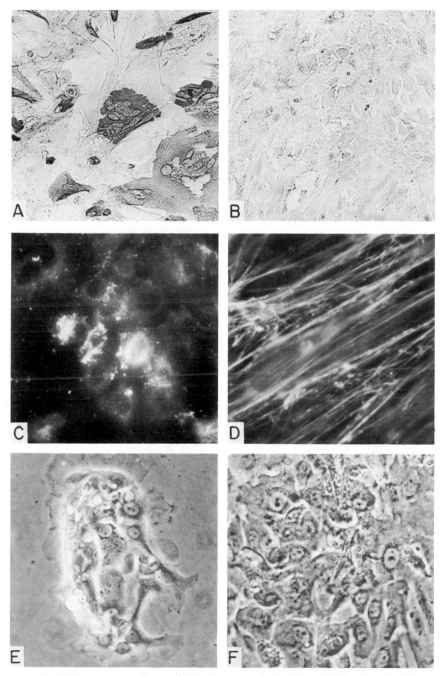

FIG. 5. (A) Immunoperoxidase staining of squamous bronchial epithelial cells with rabbit antibodies to human skin keratin. ×220. (B) Squamous epithelial cells are not stained with normal rabbit serum. ×220. (C) Immunofluorescent reaction of fibronectin in epithelial cells is restricted to the cell surface. ×350. (D) Extracellular fibronectin in bronchial fibroblasts. ×350. (E) Aggregate of epithelial cells in secondary culture. ×220. (F) Bronchial epithelial cells in tertiary culture show evidence of squamous differentiation. ×350.

Epithelial cells and fibroblasts reacted with antibody to fibronectin, however, the reaction was more intense in fibroblasts (Fig. 5C and D). In addition, the distribution of fibronectin differed in the two cell types. In epithelial cells, fibronectin was restricted to the cell surface, whereas in fibroblasts it was found on the cell surface and in fibrous material that was both cell-associated and deposited on the surface of the dish where the cells had migrated. A summary of the results obtained by immunological detection of bronchial cell products in primary cultures is given in Table III.

B. Subcultures

Attempts were made to subculture bronchial epithelial cells from primary cultures obtained from the tissues of 10 patients. In 7 cases, the cells were subcultured only once, in 2 cases, twice, and in 1 case, three times. The explants were removed from the cultures, and "contaminating" fibroblasts scraped from the plastic dishes with a rubber policeman. The most favorable conditions for the subculture of epithelial cells established to date are: (1) Cells in primary culture have to be actively growing out from the periphery of the explant at the time of subculture for appreciable growth to occur in secondary cultures. Cells from "older" primary cultures in which lateral outgrowth of epithelial cells has ceased do not appear to proliferate after subculture. (2) Cells have to be subcultured as aggregates (Fig. 5E) for subsequent growth to occur. Mitosis has not been observed in single isolated epithelial cells when the cells were subcultured into

TABLE III

PRODUCTS OF BRONCHIAL CELLS IN PRIMARY CULTURE DETECTED BY
IMMUNOLOGICAL METHODS

Product	Method[a]	Cell type	
		Epithelial	Fibroblast
A and B blood group antigens	IF	+	−
Collagen types	IF		
I		−	+
II		+[b]	−
III		−	+
IV		±	−
Keratin	IP	+	−
Fibronectin	IF	+	+

[a] IF, Indirect immunofluorescence; IP, triple layer PAP method.
[b] A small number of cells close to the area of the dish where the explant was located were positive for type II collagen, indicating that they were chondrocytes.

dishes coated with either type I or type IV collagen (Murray *et al.*, 1979) or containing irradiated (2000 R) IMR-90 human lung fibroblasts (Nichols *et al.*, 1977), or when the cells were grown in "conditioned" medium. (3) Cell survival (~60% viability as measured by the ability of the cells to exclude the dye trypan blue) and subsequent growth is best when the cells are removed from the culture dish with a prewarmed (37°C) phosphate-buffered salt solution containing glucose, polyvinylpyrrolidone, HEPES buffer, and 0.02% EGTA (solution B; Lechner *et al.*, 1978). In most experiments, enzyme solutions such as 0.05% trypsin–EDTA and 0.2% pronase are quite deleterious to the cells. The morphological appearance of bronchial epithelial cells after two subcultures is shown in Fig. 5F.

IV.   Discussion

Our laboratory is involved in the development of model systems for study of the effects of chemical carcinogens on cultured human tissues, including the bronchus. In Chapter 1, Trump *et al.* described an explant culture system for human bronchus that has been used extensively for studies of carcinogen metabolism (Harris *et al.*, 1978) and mutagenesis (Hsu *et al.*, 1978). This chapter deals with our experiences in the development of a culture system to be used, ultimately, for investigations of the mechanism(s) of carcinogen-induced cell transformation. The culture system described has been patterned, in several respects, after that developed by Marchok *et al.* (1977) for the primary culture of rat tracheal epithelial cells. Rat tracheal epithelial cells also differentiate to a squamous metaplastic epithelium in primary culture (Heckman *et al.*, 1978).

One of the first observations made in the primary culture of human bronchial cells was that the period of time the epithelial outgrowth could be maintained was dependent upon the presence of the explant. In the presence of the explant, epithelial outgrowth could be maintained for several months and the cells in the outgrowth exhibited a dynamic process of renewal, i.e., growth, differentiation, and death. However, in the absence of the explant, the epithelial cells usually underwent rapid squamous differentiation, and the majority of them sloughed into the culture medium over a period of 1–2 weeks. If fibroblasts were present in the cultures after removal of the explant, it appeared that the epithelial cells could be maintained for longer periods, suggesting that the fibroblasts produced a factor(s) that stimulated growth and/or inhibited squamous differentiation in the epithelial cells.

Another initial observation was that growth of the epithelial cells from the periphery of the explant onto the culture dish was accompanied by a rapid loss of normal mucociliary differentiation in the majority of cells. The loss of normal differentiation may have been due to several factors, among them the increased proliferation rate of the epithelial cells in the outgrowth versus those in the explant, and the loss of normal contact and interaction between the epithelial cells and the mesenchymal cells. With regard to these parameters, it should be noted that the epithelial cells in the middle portion of the explant frequently exhibited normal mucociliary differentiation at time periods in culture when the majority of the epithelial cells growing around the edge of the explant and onto the dish were undergoing squamous differentiation. In thymidine labeling studies, the labeling index of squamous epithelial cells growing along the edge of the explant was higher than that of the epithelial cells in the central portion of the explant (Barrett *et al.*, 1976).

Human bronchial epithelial cells in primary culture differentiate progressively to a squamous keratinizing epithelium. The evidence for this is as follows: (1) Squamous epithelial cells on the surface of the multilayer and exfoliated epithelial cells stain positively with the orange G–alcian blue stain for prekeratin and keratin. In addition, they stain positively by immunoperoxidase methods using antisera to prekeratin proteins isolated from human stratum corneum. (2) Numerous tonofilaments are observed in the epithelial cells by electron microscopy. (3) In recent studies, Peter Steinert (Department of Dermatology, National Cancer Institute, National Institutes of Health) isolated, by gel electrophoresis, two proteins from bronchial epithelial cells having a molecular weight and electrophoretic pattern similar but not identical to those of keratin proteins from human skin epidermal cells (unpublished data). In addition, Steinert isolated keratin polypeptides from bronchial cells that had sloughed into the culture medium. These polypeptides were assembled *in vitro* into filaments that were examined in an electron microscope (unpublished data).

Addition of β-retinyl acetate (vitamin A) to the culture medium at concentrations of 0.1, 1.0, and 5.0 μg/ml did not seem to prevent the epithelial cells from undergoing squamous differentiation as judged by light microscopy. The vitamin was toxic at a dose of 5 μg/ml as determined by the accumulation of large amounts of lipid in both the epithelial cells and fibroblasts by electron microscopy. It was not determined whether the vitamin was growth-inhibitory for the cells at this concentration, however, it has been shown to inhibit the growth of bovine pancreatic duct epithelial cells when added to the medium at a dose of 5.0 μg/ml (Stoner *et al.*, 1978).

Ciliary activity was observed in a small proportion of epithelial cells in some cultures for periods of up to 6 months *in vitro*. It was dependent upon the presence of the explant; i.e., ciliary movement usually stopped within 24–48 hours after removal of the explant from the culture, whereas in the presence of the explant it could be maintained for several months. Ciliary movement was more vigorous in epithelial cells in close proximity to the explant than in cells at a distance from the explant. This may have been because epithelial cells in close proximity to the explant grew more slowly and stratified to a greater extent than cells at a distance from the explant (Stoner *et al.*, in press). The studies of Heckman *et al.* (1978) with rat tracheal epithelial cells suggest another explanation; i.e., the number of cilia per ciliated cell may have been higher in epithelial cells in close proximity to the explant than in cells at a distance from the explant.

Reliance on phase-contrast light microscopy for cell identification can be misleading, since the morphology of specific cell types can vary under a variety of experimental conditions. Immunological and cytochemical methods can be used effectively to develop suitable markers for cell identification and to supplement morphological data. One approach to the development of immunological markers for a specific cell type is first to identify antigens unique to that cell type in tissue sections. The next step is to test cells in primary cultures for the presence of cell type-specific antigens. The cells in primary culture are usually more likely to possess the differentiated properties of their *in vivo* counterparts than cells that have been subcultured numerous times. Using this approach, we found that in both paraffin-embedded tissue sections and in primary cultures, bronchial epithelial cells from individuals with blood type A or B could be distinguished from fibroblasts using antisera to A or B blood group antigens (Katoh *et al.*, 1979). Similar results have been reported by others for other types of tissues and cells (Chessin *et al.*, 1965; Hogman, 1969). At this writing, it is not known how long cultured bronchial epithelial cells continue to possess blood group A or B antigen-specific reactivity, but the cells do react intensely after at least 90 days in primary culture, and also after at least three subcultures. In our experience, epithelial cells in both paraffin-embedded bronchial tissues and in primary cultures from patients with blood type H (O) did not react with anti-H antisera from a patient with Bombay blood group Oh (Katoh *et al.*, 1979). In addition, only epithelial cells in tissue sections from type H (O) patients reacted with anti-H lectin from *U. europaeus* (Boyd and Shapleigh, 1954); cultured epithelial cells did not react (Katoh *et al.*, 1979). Therefore, blood group antigen-specific reactivity could only be used for the identification of cultured epithelial cells from patients with blood type A or B and not those with type H (O). Blood group antigens should be suitable markers for other human epithelial cells in culture, since they are localized only in epithelial cells of many human tissues and not in stromal fibroblasts (Katoh *et al.*, 1979). Tumors of these tissues usually do not contain blood group antigens; therefore, the loss of blood group antigen-specific reactivity in normal tissues after treatment with carcinogens might be indicative of carcinogen-induced cell transformation.

Bronchial epithelial cells could also be distinguished from fibroblasts in primary cultures using antibodies to immunologically distinct collagen types. As expected from the report of Hance and Crystal (1976), fibroblasts reacted with

bronchial tissues. The technical assistance of Mr. Frank Jackson and Ms. Maria Yamaguchi, and the secretarial assistance of Ms. Maxine Bellman, is also greatly appreciated.

References

Barrett, L. A., McDowell, E. M., Frank, A. L., Harris, C. C., and Trump, B. F. (1976). *Cancer Res.* **36,** 1003–1010.

Berman, J., Stoner, G., Dawe, C., Rice, J., and Kingsbury, E. (1978). *In Vitro* **14,** 675–685.

Boren, H., Wright, E., and Harris, C. (1974). *In* "Methods in Cell Biology" (D. Prescott, ed.), Vol. 8, pp. 277–288. Academic Press, New York.

Boyd, W. C., and Shapleigh, E. (1954). *Blood* **9,** 1195–1198.

Cailleau, R.. Crocker, T. T., and Wood, D. A. (1959). *J. Natl. Cancer Inst.* **22,** 1027–1037.

Chessin, L. N., Bramson, S., Kuhns, W. J., and Hirschhorn, K. (1965). *Blood* **25,** 944–953.

Corssen, G., and Allen, C. R. (1958). *Tex. Rep. Biol. Med.* **16,** 194–202.

Engvall, E., and Rouslahti, E. (1977). *Int. J. Cancer* **20,** 1–5.

Foidart, J. M., Abe, S., Martin, G. R., Zizić, T. M., Barnett, E. V., Lawley, T. J., and Katz, S. I. (1978). *N. Eng. J. Med.* **299,** 1203–1207.

Foidart, J. M., Berman, J., Paglia, L. P., and Rennard, S. (1979). *Fed. Proc., Fed. Am. Soc. Exp. Biol.* **38,** 800/3007.

Hance, A. J., and Crystal, R. G. (1976). *In* "The Biochemical Basis of Pulmonary Function" (R. G. Crystal, ed.), Vol. II, pp. 215–271. Dekker, New York.

Harris, C. C., Autrup, H., Stoner, G. D., and Trump, B. F. (1978). *In* "Pathogenesis and Therapy of Lung Cancer" (C. C. Harris, ed.), Vol. X, pp. 559–607. Dekker, New York.

Heckman, C. A., Marchok, A. C., and Nettesheim, P. (1978). *J. Cell Sci.* **32,** 269–291.

Hoch-Ligeti, C., and Hobbs, J. P. (1958). *Proc. Soc. Exp. Biol. Med.* **97,** 59–62.

Hogman, C. F. (1969). *Exp. Cell Res.* **21,** 137–143.

Hsu, I. C., Stoner, G. D., Autrup, H., Trump, B. F., Selkirk, J. K., and Harris, C. C. (1978). *Proc. Natl. Acad. Sci. U.S.A.* **75,** 2003–2007.

Katoh, Y., Stoner, G. D., McIntire, K. R., Hill, T. A., Anthony, R., McDowell, E. M., Trump, B. F., and Harris, C. C. (1979). *J. Natl. Cancer Inst.* **62,** 1177–1185.

Lechner, J. F., Narayan, K. S., Ohnuki, Y., Babcock, M. S., Jones, L. W., and Kaighn, M. E. (1978). *J. Natl. Cancer Inst.* **60,** 797–801.

Lee, L. D., Baden, H. P., Kubilus, J., and Fleming, B. F. (1976). *J. Invest. Dermatol.* **67,** 521–525.

Leibovitz, A. (1963). *Am. J. Hyg.* **78,** 173–180.

McDowell, E. M., and Trump, B. F. (1976). *Arch. Pathol. Lab. Med.* **100,** 405–414.

Marchok, A. C., Rhoton, J. C., Griesemer, R. A., and Nettesheim, P. (1977). *Cancer Res.* **37,** 1811–1821.

Miller, E. J., and Matukas, V. J. (1974). *Fed. Proc., Fed. Am. Soc. Exp. Biol.* **33,** 1197–1204.

Murray, J. C., Stingl, G., Kleinman, H. K., Martin, G. R., and Katz, S. I. (1979). *J. Cell Biol.* **80,** 197–203.

Nichols, W. W., Murphy, D. G., Cristofalo, V. J., Toji, L. H., Greene, A. E., and Dwight, S. A. (1977). *Science* **196,** 60–63.

Nowack, H., Gay, S., Wick, G., Becker, V., and Timpl, R. (1976). *J. Immunol. Methods* **12,** 117–124.

Steele, V. E., Marchok, A. C., and Nettesheim, P. (1977). *Int. J. Cancer* **20,** 234–238.

Sternberger, L. A., Hardy, P. H., Cuculis, J. J., and Meyer, H. G. (1970). *J. Histochem. Cytochem.* **18,** 315–333.

Stoner, G. D., Harris, C. C., Bostwick, D. G., Jones, R. T., Trump, B. F., Kingsbury, E. W., Fineman, E., and Newkirk, C. (1978). *In Vitro* **14,** 581–589.

Stoner, G. D., Harris, C. C., Myers, G. A., Trump, B. F., and Connor, R. D. *In Vitro* (in press).

Suemasu, K., and Mizuta, T. (1975). *Gann* **66,** 109–110.

Trump, B. F., Valigorsky, J. M., Dees, J. H., Mergner, W. J., Kim, K. M., Jones, R. T., Pendergrass, R. E., Garbens, J., and Cowley, R. A. (1974). *Hum. Pathol.* **4,** 89–109.

Yaoita, H., Foidart, J. M., and Katz, S. I. (1978). *J. Invest. Dermatol.* **70,** 191–193.

Chapter 3

The Fibroblast of Human Lung Alveolar Structures: A Differentiated Cell with a Major Role in Lung Structure and Function

KATHRYN H. BRADLEY, OICHI KAWANAMI, VICTOR J. FERRANS AND RONALD G. CRYSTAL

Pulmonary Branch and Pathology Branch,
National Heart, Lung, and Blood Institute,
National Institutes of Health,
Bethesda, Maryland

ISBN 0-12-564121-4

I.   Introduction

Human lung fibroblasts have been widely studied since the description of the Wistar Institute series of cell strains by Hayflick and Moorhead (1961). In general, the majority of these studies have not focused on the fibroblast as a lung cell, but rather have used it for investigations of viral growth and vaccine production (Hayflick *et al.*, 1962), as a model system for the study of aging *in vitro* (Hayflick, 1977) and for various metabolic studies (Cristofalo, 1972). In contrast, the purpose of this chapter is to evaluate the human lung fibroblast as a lung cell that plays a major role in lung structure and function. Specifically, we will focus on this cell in terms of (1) its derivation from, and relationship to, other cells of the alveolar structures; (2) its general properties; (3) its specific functions within the lung; (4) the optimal methodology for its culture from fetal and adult human lung; (5) its morphology in culture; (6) its ability to maintain the differentiated state *in vitro;* and (7) its relationship to fibroblasts of other organs.

The postnatal lung is a complex organ with one major function: to mediate the exchange of oxygen and carbon dioxide between the atmosphere and the blood. To accomplish this, the lung brings air and blood into close proximity in structures termed "alveoli." In the adult human lung, alveoli are comprised of four major cell types (Weibel *et al.*, 1976; Kuhn, 1976):[1] type I epithelial cells, type II epithelial cells, endothelial cells, and mesenchymal cells (Fig. 1). Mesenchymal cells are found entirely in the region of the alveolar structures termed the "interstitium," a region whose boundaries are defined by the basement membranes of alveolar epithelial and endothelial cells (Weinberger and Crystal, 1979). Careful morphometric studies of normal human lung have demonstrated that the alveolar structures of both lungs contain a total of 230×10^9 cells. These alveolar cells are present in the following proportions: type I epithelial, 8%; type II epithelial, 16%; endothelial, 30%; and mesenchymal cell, 37%; the remaining 9% of the cells are inflammatory and immune effector cells, predominately macrophages (Barry *et al.*, 1979). Thus, mesenchymal cells constitute almost two-fifths (approximately 84×10^9 cells) of all cells comprising the gas-exchanging structures of the lung.

A.   Derivation and Definition of the Alveolar Fibroblast

Like the endothelial cell, the alveolar mesenchymal cell is derived from the primitive mesoderm (Burri and Weibel, 1977). During the canalicular stage

---

[1]Inflammatory and immune effector cells such as alveolar macrophages and lymphocytes are also present within the alveolar structures. Although these cells play major roles in lung structure, function, and defense, they are not generally considered permanent cells of the alveoli. The human lung macrophage is discussed in detail by Hunninghake *et al.* in Chapter 6.

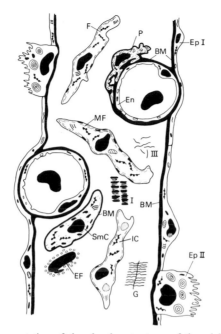

Fig. 1.   Schematic representation of the alveolar structures of the adult human lung. On the epithelial surface there are two types of cells: epithelial type I (EpI) and epithelial type II (EpII). Capillaries are lined with endothelial cells (En). Basement membrane (BM) lies beneath the epithelial and endothelial cells. Five types of mesenchymal cells are shown within the alveolar interstitium: fibroblasts (F), myofibroblasts (MF), interstitial cells (IC), pericytes (P), and smooth muscle cells (SmC). Note that pericytes and smooth muscle cells are surrounded by basement membrane, whereas fibroblasts, myofibroblasts, and interstitial cells are not. Also present within the interstitium are connective tissue components including type I collagen (I), type III collagen (III), elastic fibers (EF), and proteoglycans (G). Other connective tissue constituents such as fibronectin and type IV and type V collagen are not shown. Also omitted are alveolar macrophages and lymphocytes, cells that comprise the normal inflammatory and immune effector cells of the alveolar structures.

(16–24 weeks) of human fetal lung development, cells destined to become alveolar mesenchymal cells begin to form an interstitial mesenchyme. The details of human alveolar mesenchymal cell ontogeny are not known, but in the human lung several morphological "types" of mesenchymal cells have been described in the alveolar interstitium (Table I). Different investigators have described each mesenchymal cell type in some detail, but little attention has been given to the possible relationships among these cells. The most important *in situ* morphological characteristics of the five different types of alveolar mesenchymal cells that have been described are as follows:

1. The "fibroblast" is generally regarded as an elongated cell with numerous long cytoplasmic processes and an oval nucleus with smooth contours, often containing a nucleolus (Ross, 1968a,b). This cell usually has abundant rough

TABLE I

Morphological Features of Mesenchymal Cells Identified within the Alveolar Interstitium of Normal Lung

Cell type	Basement membrane	Cytoplasmic branching	Pinocytotic vesicles	Rough endoplasmic reticulum	Free ribosomes	Lipid droplets	Actin-like cytoplasmic filaments	Peripherally located dense bodies
Fibroblast	0	Moderate	0	Abundant	Moderate	0	Few	Few or none
Interstitial cell	0	Extensive	0	Variable	Few	+	Moderate	Moderate
Myofibroblast	0	Moderate to extensive	0	Abundant	Few	0	Moderate	Few to moderate
Pericyte	+	Extensive	+	Sparse	Few	0	Few	Few
Smooth muscle cell	+	Minimal	+	Sparse	Few	0	Abundant	Abundant

endoplasmic reticulum and free ribosomes, a well-developed Golgi complex, a smooth plasma membrane, and small amounts of filaments, 30–80 Å in diameter, adjacent and parallel to the plasma membrane (Brandes *et al.*, 1972). It is not surrounded by basement membrane.

2. The "interstitial cell" (also called the "interstitial connective tissue cell") is described as an elongated cell with long cytoplasmic extensions that often are partially wrapped around capillaries (Kuhn, 1976). These cells have variable amounts of endoplasmic reticulum and Golgi components and are characterized by large cytoplasmic lipid droplets (Kuhn, 1976; Brody and Vaccaro, 1979; Vaccaro and Brody, 1978). The nature of these lipid droplets is not clear, but animal studies have suggested that they contain esterase activity against various artificial substrates (O'Hare *et al.*, 1971). At least part of the lipid contained within these cytoplasmic structures is synthesized by the cell, as autoradiographic studies with adult rats administered labeled palmitate have shown that the label appears in the lipid droplets within 10 minutes (Askin and Kuhn, 1971).

3. The "myofibroblast" (also called a "contractile cell") is an alveolar mesenchymal cell with an abundance of peripherally located actin-like 30 to 80-Å cytoplasmic filaments running parallel to (and often attaching to) the plasma membrane (Kapanci *et al.*, 1974; Kapanci and Costabella, 1979). The nuclei of myofibroblasts show indentations and folds that reflect the degree of contraction of these cells (Ryan *et al.*, 1973). Since lung strips contract in response to various stimuli, and since the myofibroblast contains cytoplasmic structures which might mediate such contraction, it has been hypothesized that the myofibroblast plays a role in regulating air and blood relationships by modulating alveolar geometry (Kapanci *et al.*, 1974).

4. The "pericyte" is defined by its close relationship to capillary endothelial cells (Rhodin, 1968). The pericyte is ensheathed by the endothelial basement membrane, contains branched cytoplasmic processes with fine, actin-like filaments and some microtubules, has pinocytotic vesicles on its surface closest to the endothelial cell, and contains few lysosomes. In general, the larger cytoplasmic processes of the pericyte are separate from the endothelial cell, but the finer processes penetrate through the basement membrane and establish close contact with the endothelium (Weibel, 1974).

5. Smooth muscle cells are rarely found in the interstitium of normal human alveolar structures (Kapanci *et al.*, 1974). When present, they are usually observed in association with the vasculature or at the opening to a group of alveoli (i.e., alveolar ducts). In general, they have the same characteristics as smooth muscle cells elsewhere in the body. Most importantly, alveolar smooth muscle cells contain thin, actin-like filaments throughout their cytoplasm and peripherally located dense bodies which serve as attachment sites for the filaments; these cells are surrounded by a thin basement membrane (Kuhn, 1976). Like myofibroblasts, nuclei of smooth muscle cells have folds reflecting their degree of contraction (Lane, 1965; Bloom and Cancilla, 1969).

It is likely that these five types of alveolar mesenchymal cells form a continuum, although the order and polarity (if any) of differentiation among them is unclear (Franks and Cooper, 1972; Fritz *et al.*, 1970; Min, 1979). However, according to the morphological criteria given, these cells can be grouped into two general populations: the fibroblast, interstitial cell, and myofibroblast group; and the pericyte and smooth muscle cell group. The justification for each grouping is as follows:

1. None of the cells of the fibroblast group are associated with basement membrane, while both pericytes and smooth muscle cells are completely surrounded by basement membranes.

2. Pericytes and smooth muscle cells have prominent pinocytotic vesicles, while such surface structures are rarely observed in cells of the fibroblast group.

3. While all five cell types have 30- to 80-Å cytoplasmic filaments, these filaments are extensive in pericytes and smooth muscle cells but confined to the cell periphery in the other cell types. Although such filaments are a major morphological feature of myofibroblasts, they are also found *in situ* in fibroblasts and interstitial cells, albeit to a lesser degree. Thus, while there are differences among the five cell types in terms of cytoplasmic filaments, the differences are minor within the fibroblast group yet major between the fibroblast group and the pericyte–smooth muscle cell group.

4. Although cytoplasmic lipid droplets may give the interstitial cell a distinctive appearance, there is no known function of these droplets to suggest they impart to the cell a specific capability different from those of the fibroblast and myofibroblast. Thus, unless information accrues to the contrary, it is reasonable to assume that the formation of intracytoplasmic lipid droplets is simply a variation in the general properties of the cells of the fibroblast group.

As will be described in detail, since the morphological and functional features of cultured human lung mesenchymal cells are generally similar to those of the first group, for convenience we will use the term "fibroblasts" to refer to the population of cells to be discussed in this chapter. Since the donor tissues for human lung fibroblasts are fragments of lung parenchyma (i.e., the distal lung containing primarily alveoli and devoid of large arteries, veins, and bronchi), it is reasonable to assume that the cells called "lung fibroblast" (WI-38, MRC-5, IMR-90, HFL1, etc.) are likely to be derived from alveolar structures. Thus, we will use the terms "lung fibroblast" and "alveolar fibroblast" interchangeably. To our knowledge, human lung pericytes have never been maintained in culture, nor have human alveolar smooth muscle cells.

B. General Properties of Human Alveolar Fibroblasts

Since the human fetal lung fibroblast is widely used as an *in vitro* model of the aging process, as well as in studies of viral transformation, a large amount of data has accrued concerning its general metabolic properties. In general, most of this information has been derived from studies of the cell lines WI-38, MRC-5, and IMR-90, all of which were cultured from the alveolar structures of 13- to 18-week-old human fetal lungs (American Type Culture Collection, 1979). As expected, these cells have all the metabolic machinery for active DNA (Macieira-Coelho et al., 1966b), RNA (Macieira-Coelho et al., 1966a), and protein (Sinclair and Leslie, 1959) synthesis. They contain all the necessary enzymes for glucose metabolism (Holliday and Tarrant, 1972; Wang et al., 1970; Bartholomew et al., 1969), including the pentose phosphate shunt (Cristofalo, 1970); under usual culture conditions they utilize 3.34 μmoles glucose per 10^6 cells per day (Cristofalo and Kritchevsky, 1965). Interestingly, lung fibroblasts produce lactate in a 2:1 ratio relative to the amount of glucose consumed, even though they contain the enzymes of the Krebs cycle (Wang et al., 1970) and are exposed to adequate amounts of oxygen (Cristofalo and Kritchevsky, 1965, 1966). The reason for this is unclear, but it may be one reason why the lung has a net production of lactate (Tierney and Levy, 1976) even though it is the most aerobic organ in the body. Although glucose is the most effective substrate for the glycolytic pathway in lung fibroblasts, other sugars can be utilized, of which mannose is the most effective (Cristofalo and Kritchevsky, 1965). Lung fibroblasts can also store glycogen; under resting conditions WI-38 cells have 9.7 μg glycogen per 10^6 cells (Cristofalo and Kritchevsky, 1966).

In addition to metabolizing glucose, human fibroblasts synthesize lipids, including cholesterol (Rothblat et al., 1971). These cells also contain a variety of lysosomal acid hydrolases (Wang et al., 1970; Bartholomew et al., 1969; Turk and Milo, 1974; Bosmann et al., 1976), as well as acid and neutral proteases (Bosmann et al., 1976), likely used in general metabolic processes.

C. Specific Functions of the Alveolar Fibroblast (Table II)

In vivo, the alveolar fibroblast directly or indirectly influences overall lung structure and function in at least three ways: (1) directly because of its inherent properties, (2) indirectly by maintaining interstitial connective tissues, and (3) indirectly by defending the lung.

The fibroblast directly influences alveolar structure simply by its abundance; in the adult lung it occupies approximately 75% of the total volume of the interstitium (Weibel et al., 1976). In addition, it likely plays an active role in regulating air and blood flow at the alveolar level through its contractile elements (i.e., one of the roles of the so-called myofibroblast) (Kapanci et al., 1974).

Indirectly, the fibroblast significantly influences alveolar structure and function through its major role in maintaining interstitial connective tissue. Of the five classes of connective tissue found within the alveolar interstitium (interstitial collagens, basement membrane components, elastic fibers, proteoglycans, and fibronectin) (Hance and Crystal, 1975; Bray, 1978), the fibroblast produces at

TABLE II

FUNCTIONS OF THE ALVEOLAR FIBROBLAST

Function	Reference
Direct influence on structure and function	
Influences structure because it occupies a large proportion of interstitial volume	Weibel *et al.*, 1976
Contractile properties	Kapanci *et al.*, 1976
Maintenance of interstitial connective tissue	
Collagen production	Hance *et al.*, 1976; Breul *et al.*, 1979
Secretion of collagenase	Horwitz *et al.*, 1977
Phagocytosis of collagen	Rose *et al.*, 1978
Proteoglycan production	Sjöberg and Fransson, 1977
Fibronectin production	Baum *et al.*, 1977
Lung defense	
Phagocytosis, pinocytosis	Henderson *et al.*, 1975
Liberation of chemotactic factors	Sobel and Gallin, 1979
Fibrinolysis	Mott *et al.*, 1974
Production of complement components	Al-Adnani and McGee, 1976; Reid and Solomon, 1977
Secretion of antioxidants	Yamanaka and Deamer, 1974
Secretion of protaglandins	Taylor and Polgar, 1977
Uptake and degradation of α2-macroglobulin	Van Leuven *et al.*, 1978
Synthesis and secretion of α2-macroglobulin	Mosher *et al.*, 1977; Mosher and Wing, 1976
Production of interferon	Kronenberg, 1977

least three: collagen, proteoglycans, and fibronectin. The predominant interstitial collagens are of two types: type I (thick, cross-banded fibers) and type III (thin, randomly dispersed fibrils). The human alveolar fibroblast not only produces both collagen types (Hance *et al.*, 1976), but also secretes a collagenase, an enzyme that, when activated, can destroy both collagens I and III (Kelman *et al.*, 1977). Morphological studies have demonstrated that human fibroblasts are also capable of ingesting collagen fibrils and thus use phagocytotic mechanisms to modulate the connective tissue matrix (Yajima and Rose, 1977; Rose *et al.*, 1978; Eley and Harrison, 1975; Ten Cate *et al.*, 1976). Although production of the protein moiety of proteoglycans has not been evaluated in cultured lung fibroblasts, it is known that these cells synthesize the major glycosaminoglycans of proteoglycans (Sjöberg and Fransson, 1977). Of all the proteins secreted by lung fibroblasts, fibronectin is second in abundance only to collagens (Baum *et al.*, 1977). This macromolecule is found on the surface of alveolar fibroblasts, as well as in the interstitial connective tissue and basement membrane (Bray, 1978).

As in other tissues, fibronectin in the alveolar structures is thought to modulate cell–matrix and cell–cell interactions and thus may be an important determinant of cellular order, particularly in relationship to the interstitial connective tissue matrix (Hedman *et al.*, 1978; McDonald *et al.*, 1979).

Although not as well defined as its role in connective tissue maintenance, there is increasing evidence that the alveolar fibroblast contributes to lung defense. Human lung fibroblasts have not been evaluated in detail, but there is ample evidence that fibroblasts are capable of ingesting proteins and various particulates and thus may form a secondary line of defense against foreign materials (Steinman *et al.*, 1974; Henderson *et al.*, 1975). Fibroblasts seem to accomplish this by at least two mechanisms: pinocytosis for protein ingestion (Steinman *et al.*, 1974), and multibridge interactions for particulates (e.g., via lectins) (Goldman, 1977). Interestingly, although fibroblasts seem to be less active than macrophages in pinocytosis, it has been suggested that they are much larger cells and therefore, on a per cell basis, may contribute as much as macrophages in this form of lung defense (Steinman *et al.*, 1974). Recently, it has been suggested that fibroblasts recruit other cells in this defense role by the production of chemotactic factors that attract circulating neutrophils (Sobel and Gallin, 1979). Fibroblasts may also contribute to the inflammatory response by secreting prostaglandins (Dayer *et al.*, 1977; Baenziger *et al.*, 1977; Mathé *et al.*, 1977). This function has not been well studied in human lung fibroblasts, but it is known that these cells are capable of producing prostaglandins (Taylor and Polgar, 1977). Although controversial, it has been suggested that the alveolar fibroblast may play an important role in protecting the lung against proteolytic damage by synthesizing and secreting $\alpha 2$-macroglobulin, a major, general-purpose antiprotease (Mosher and Wing, 1976; Mosher *et al.*, 1977; Van Leuven *et al.*, 1978). Interestingly, various strains of human lung fibroblasts also have been shown to secrete at least one active protease, plasminogen activator; this enzyme likely plays a role in clearing the alveolar structures of fibrin debris (Pearlstein *et al.*, 1976; Mott *et al.*, 1974; Bernik and Kwaan, 1969). Compared to other cell strains, lung fibroblasts are regarded as very active producers of plasminogen activator (Bernik and Kwaan, 1969) and thus, along with alveolar macrophages, these cells may significantly contribute to fibrinolysis. There are probably internal controls on this system, as alveolar fibroblasts also produce an inhibitor of this protease (Bernik and Kwaan, 1969). It is also clear that the fibroblast contributes to the components of the classic complement system, as these cells actively synthesize and secrete Clq, Clr, and Cls (Al-Adnani and McGee, 1976; Reid and Solomon, 1977). This is consistent with the finding that C4, a later component of the classic complement system, is found in human alveoli (Reynolds *et al.*, 1977). It also appears that lung fibroblasts may protect the lower respiratory tract from oxidant injury by their production of superoxide dismutase, an enzyme that catalyzes the conversion of superoxide radicals to hydrogen peroxide and thus helps to prevent peroxidative damage to alveolar structures (Yamanaka and Deamer, 1974).

II. Materials and Methods

A. Culture of Fetal Lung Fibroblasts: Enzymatic Dispersal Method

Lungs were obtained from human fetuses following abortion and stored briefly in a sterile culture medium until processing could be initiated. Various media, such as Ham's F12, Dulbecco's modified Eagle's, medium 199, and RPMI 1640, all supplemented with 10% fetal calf serum (Colorado Serum Company, Denver, Colorado), penicillin (100 U/ml), and streptomycin (100 μg/ml) have been found equally satisfactory for initiating and growing lung fibroblasts. The tissue (1-3 gm) was minced finely with scissors, washed twice with phosphate-buffered saline (PBS), and once with 0.25% trypsin in PBS (without calcium or magnesium; Gibco, Grand Island, New York). The tissue fragments were incubated at 37°C with 10- to 15-ml portions of fresh trypsin solution for 20 minutes, the tissue was allowed to settle, and the supernatant (containing dispersed cells) was decanted into a capped tube containing equal volumes of culture medium and 10% fetal calf serum. The residual tissue was incubated twice more with 10-ml portions of fresh trypsin solution, and the detached cells separated by decanting. After the third digestion, practically all the remaining tissue could be dispersed by drawing up and down several times in a 10-ml pipet. The entire pool of cells was then filtered through three to four layers of sterile gauze and centrifuged at 400 g to pellet the cells. The pellet was resuspended in fresh medium, and a count made of viable cells. Cultures were set up at 6-10 $\times$ 10^6 cells per 75-cm^2 Falcon flask or per 100-mm petri dish (Fig. 2). The primary cultures were placed in a 37°C humidified incubator gassed with the required percentage of carbon dioxide-air for the medium being used. In general, the primary cultures were ready for subculturing in less than a week and thereafter were routinely subcultured every 5 days using a 1:3 split ratio.

B. Culture of Adult Lung Fibroblasts: Enzymatic Dispersal Method

Biopsies of peripheral adult human lung were cultured by the same procedure used for fetal lung fibroblasts. However, the tissue fragments of adult lung floated in the trypsin solution, and thus it was necessary to harvest the dispersed cells by pipetting rather than decanting. The yield of dispersed cells from adult

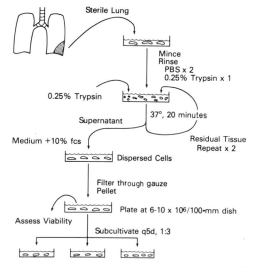

Sterile Lung

Mince
Rinse
PBS x 2
0.25% Trypsin x 1

0.25% Trypsin

Supernatant

37°, 20 minutes

Medium +10% fcs

Residual Tissue
Repeat x 2

Dispersed Cells

Filter through gauze
Pellet

Plate at 6-10 x 10⁶/100-mm dish

Assess Viability

Subcultivate q5d, 1:3

FIG. 2. Methods used to culture human alveolar fibroblasts. Explant methods can also be used, but the enzymatic dispersal technique shown here is faster and easier (see Section II for details). fcs, Fetal calf serum.

lung was considerably reduced compared to that from fetal lung (per milligram wet weight of tissue), even with longer periods of trypsin digestion. There was little improvement in yield with the use of a mixture of trypsin and collagenase (Worthington Biochemical Corporation, Freehold, New Jersey), or trypsin and hyaluronidase (Worthington). Dispersed adult lung cells were plated at the same high cell density as fetal cells, although this often meant having only one or two culture flasks initially. In some cases cell growth was quite slow. In these instances, cultures were fed every 5 days until there was sufficient growth for the first subcultivation, sometimes a period of up to 3 weeks. The primary cultures of adult lung obtained by this method contained a heterogeneous population of cells, but by subcultivation five to eight consisted entirely of a population of fibroblasts.

C. Culture of Fetal or Adult Fibroblasts: Outgrowth Method

Cultures of fetal or adult lung fibroblasts could be obtained without resorting to the use of enzymatic digestion, i.e., by using outgrowth of lung explants. However, in the case of adult or newborn lung, floating of the tissue became a problem. The following method has been found helpful in securing attachment of the fragments to the culture surface, a necessary condition for the outgrowth of cells.

Lung tissue was minced with scissors into 1- to 2-mm³ pieces and rinsed three

TABLE III

GENERAL CHARACTERISTICS OF THE CULTURED
HUMAN ALVEOLAR FIBROBLAST

Parameter	Range
Doubling time	20–40 hours
Plating efficiency	5–25%
Life span	40–60 divisions
Cell cycle	
S	6–10 hours
G_2	4–6 hours
G_1	4–15 hours
DNA per cell	9–10 pg
Mitotic index	
Log	1.8
Confluency	0.4
[^{3}H]thymidine uptake per cell	
Log	51.2 dpm
Confluency	2.2 dpm

alveolar structures vastly dominated the specimens, and thus it is reasonable to assume that the cultured fibroblasts were probably of alveolar origin. The primary cultures were heterogeneous, with islands of polygonal epithelial-like cells surrounded by fibroblasts, but by the fifth or sixth subcultivation the cultures were entirely fibroblastic in appearance.

In approximately one-third of the cases, the source tissue was normal lung; in the remainder the donor had a lung disorder, most commonly an interstitial lung disease such as idiopathic pulmonary fibrosis or sarcoidosis (Fulmer and Crystal, 1979). Five of these cell strains are maintained and distributed by the American Type Culture Collection. The most extensively studied is HFL-1 (ATCC CLL-153), a fibroblast strain derived from a 16-week-old human fetal male. The

FIG. 3 and 4.   Phase-contrast (Fig. 3) and Nomarski differential interference contrast (Fig. 4) micrographs of a cultured 5-year-old human alveolar fibroblast (strain L-24) in the logarithmic phase of growth. The cell is elongated and has a centrally located nucleus, and its cytoplasmic processes extend in several directions. Mitochondria and other cytoplasmic organelles are clustered about the nucleus. ×525.

FIG. 5.   High-magnification micrograph, taken with the Nomarski system, showing the smooth appearance of the surface of a cultured fetal human alveolar fibroblast (strain HFL-1). Compare with scanning electron micrograph in Fig. 5. ×1800.

FIG. 6.   Micrograph taken as in Fig. 5, showing surface ridges and microvilli in a cultured 5-year-old human alveolar fibroblast (strain L-24). Compare with scanning electron micrograph in Fig. 8. ×1800.

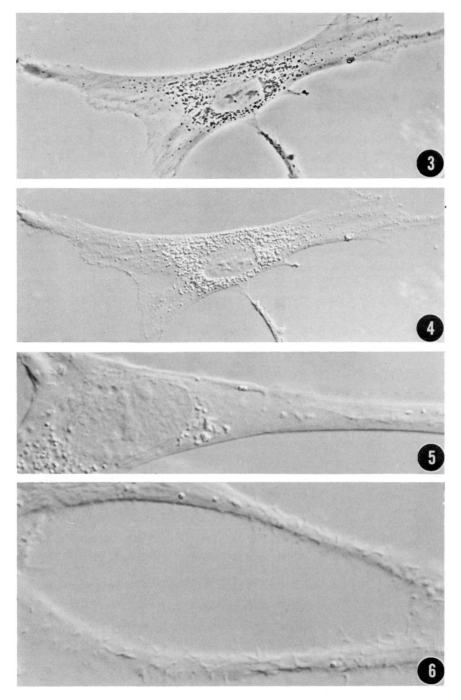

others include CCL-151, derived from a 5-year-old normal male, CCL-135 from a 16-year-old normal male, CCL-134 from a 26-year-old female with idiopathic pulmonary fibrosis, and CCL-191 from a 48-year-old female with idiopathic pulmonary fibrosis.

Karyology of all cell strains demonstrated a normal karyotype at all periods studied. The DNA/cell ratio of the human lung fibroblasts was 9.2 ± 0.41 pg/cell, similar to that described for other human cells (Fujimoto *et al.*, 1977; Cristofalo and Kritchevsky, 1969; Schneider and Shorr, 1975). The cell strains have been periodically evaluated for mycoplasma and have always been negative. Morphological evaluation for viral particles using transmission electron microscopy has also been negative.

With conventional methods (Schneider and Shorr, 1975), estimation of the doubling time of these cell strains revealed a range of 20–40 hours (Table III). In general, fibroblasts from younger donors tended to grow faster, but this was not always the case. HFL-1, the fetal fibroblast strain, showed evidence of senescence after subcultivations 33–38 (50–57 population doublings). In comparison, the number of population doublings of the adult strains was variable, but always less than that of the fetal strain. In general, the onset of senescence was indicated by an increase in the doubling time to 40 hours or more, the appearance of increasing numbers of unusually large cells, the accumulation of debris which appeared to be on the surface of the cell layer, and breaks in the evenness of the cell sheet.

The plating efficiency of the human lung fibroblast cell strains varied from 5 to 25% (Table III); cells derived from younger donors tended to have a lower plating efficiency. Studies by M. Schafer in our laboratory have shown that, in general, times for cell cycle subdivisions of human lung fibroblasts are similar to those described in the literature for diploid human cells (Grove and Cristofalo, 1976).

B.   Morphology

Soon after subculturing, pulmonary fibroblasts spread out and appear polygonal or stellate-shaped, with elongated cytoplasmic processes that extend in various directions (Figs. 3–6). As the cells achieve confluence, they become spindle-shaped and assume the well-known parallel arrangement of fibroblasts in tissue culture. The surfaces of pulmonary fibroblasts in tissue culture generally

---

FIG. 7.   Scanning electron micrograph of a cultured fetal human alveolar fibroblast (strain HFL-1), showing the smooth appearance of its surface and long, slender cytoplasmic processes. ×5000.

FIG. 8.   Scanning electron micrograph of a cultured 5-year-old human alveolar fibroblast (strain L-24), showing ridges and microvillous projections from its surface. ×7000.

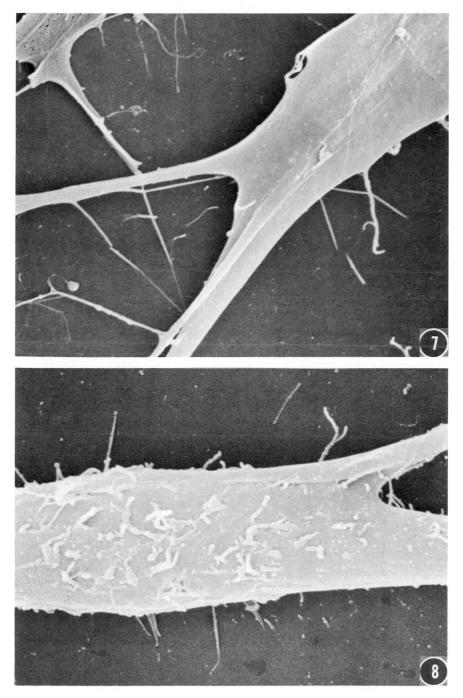

appear smooth on scanning electron microscopic examination; only a few microvilli are present (Fig. 7). However, there are heterogeneities in the surface
appearance of human lung fibroblasts; the ATCC CCL-151 strain of fibroblasts
shows small but prominent longitudinally oriented ridges on the surfaces of
cytoplasmic processes (Fig. 8). These ridges appear to be associated with the
shape transition from stellate to spindle-like. They are visible by *in situ* light
microscopic examination using the Nomarski differential interference contrast
technique (Fig. 6) and also by scanning electron microscopic examination of
cells fixed *in situ* (Fig. 8). Small amounts of finely filamentous material appear
to be attached to some areas of the surfaces of cells in the stationary phase of
growth, but no well-developed basal lamina is seen.

After becoming confluent, pulmonary fibroblasts in tissue culture measure up
to 100 μm in length and 10 μm in maximal width. Their nuclei are elongated and
contain prominent nucleoli (Fig. 9). The chromatin is evenly distributed, with
only minimal margination. The contours of the nuclear membranes are smooth.
The cytoplasm contains the Golgi complex, mitochondria, free ribosomes, cisterns of rough-surfaced endoplasmic reticulum (Fig. 10), lysosomes, myelin figures and residual bodies (Fig. 11), vacuoles with a clear content, microtubules,
microfilaments (Figs. 10 and 12), and occasional coated vesicles; pinocytotic
vesicles are rare. The Golgi complex is of moderate size and consists of
smooth-walled cisterns associated with vesicles. The mitochondria are small and
elongated. The free ribosomes often occur in clusters rather than singly. The
cisterns of rough-surfaced endoplasmic reticulum are elongated, are often arranged in parallel, and contain variable amounts of a finely dispersed material of
low electron density (Fig. 10). Lysosomes and residual bodies of lysosomal
origin are abundant (Figs. 9 and 11). The morphology of the last-mentioned
structures is variable and complex, ranging from clear vacuoles to large arrays of
concentric lamellae limited by single membranes and associated with electron-
dense particles 300–600 Å in diameter. The nature of such particles is unknown.

Fig. 9. Transmission electron micrograph of cultured fetal human alveolar fibroblasts (strain
HFL-1) in the stationary phase of growth. Note the centrally located, elongated nucleus, the small,
sparse mitochondria, the cisterns of rough-surfaced endoplasmic reticulum, and the moderate number
of residual bodies. Small amounts of filamentous material are adjacent to several areas of the cell
surfaces. ×11,500.

Fig. 10. Portions of two adjacent cultured fetal human alveolar fibroblasts (strain HFL-1) in the
stationary phase of growth. Note the 100-Å microfilaments (arrowheads) in the cell at the top and the
parallel cisterns of rough-surfaced endoplasmic reticulum in the cell at the bottom. ×44,000.

Fig. 11. Residual bodies in cultured fetal human alveolar fibroblasts (strain HFL-1) in the
stationary phase of growth are composed of small particles and concentric, electron-dense lamellae.
×34,000.

Fig. 12. Actin-like microfilaments form a narrow, peripherally located band in cultured human
alveolar fibroblasts (strain HFL-1). ×55,000.

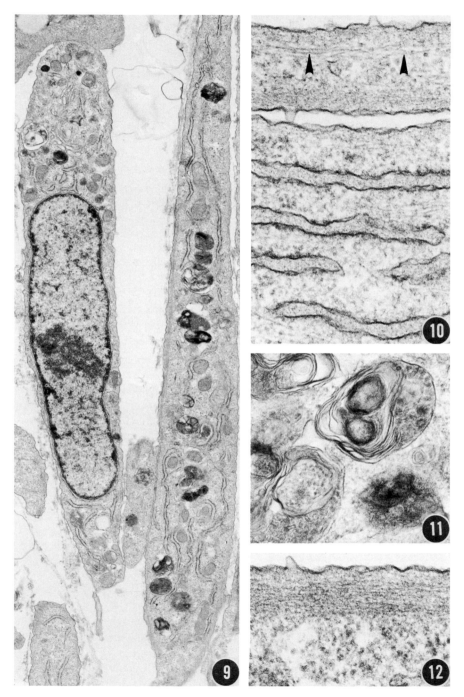

Two types of microfilaments are present in the cytoplasm. The larger of these filaments (Fig. 10) measure 100 Å, are presumed to function as a cytoskeleton (Weihing, 1979), and form small, discrete bundles in the cytoplasm. The smaller ones (Fig. 12) measure from 50 to 70 Å in diameter, are distributed mainly along the subplasmalemmal regions of the cells, and correspond to the actin-containing filaments of other cell types (Lazarides, 1976).

Cells in the logarithmic phase of growth (Fig. 13) differ from cells in the stationary phase in that they have numerous free ribosomes (often arranged in clusters or polyribosomes) and less numerous cisterns of rough-surfaced endoplasmic reticulum (these are narrower and less filled with material presumed to be newly synthesized protein). The most striking feature of older, confluent fibroblasts is the presence of large numbers of pleomorphic residual bodies.

C. Stability of Phenotypic Expression of Cultured Human Alveolar Fibroblasts

Of the many functions of the alveolar fibroblast, the production of collagen is one of the most important. Collagen is the most abundant extracellular protein of the alveolar structures, and it plays a major role in defining alveolar structure and function. The normal human alveolar interstitium contains at least four collagen types (Crystal *et al.*, 1978): the major types (in terms of quantity) are interstitial collagens (types I and III), and the minor types are basement membrane collagens (types IV and V) (Madri and Furthmayr, 1979).

For the cultured human alveolar fibroblast to be of value as in an *in vitro* model of *in vivo* function, it is critical that its phenotypic expression with respect to a major function such as collagen production be maintained *in vitro*. This is not a trivial question, as recent studies have demonstrated that the magnitude of collagen production by fibroblasts may be very sensitive to their environmental milieu, such that under typical culture conditions they produce less collagen than *in vivo* (Schwarz and Bissell, 1977). There are several lines of evidence, however, indicating that cultured human alveolar fibroblasts fully maintain their differentiated state *in vitro* with respect to collagen production.

1. In culture, the human alveolar fibroblast synthesizes and secretes both collagens I and III (Hance *et al.*, 1976; Kelley *et al.*, 1979), but it synthesizes more type I than type III. This observation is consistent with data indicating that (a) although alveolar epithelial and endothelial cells likely produce collagen (Crystal *et al.*, 1978), in terms of numbers, fibroblasts are the major collagen-producing cells of the alveolar structures; and (b) collagen types I and III are the major interstitial collagens (>90% of the total), with type I predominating (Hance *et al.*, 1976).

2. Cultured human alveolar fibroblasts produce approximately 6×10^5 collagen chains per hour per cell (Breul *et al.*, 1979). In comparison, explants of

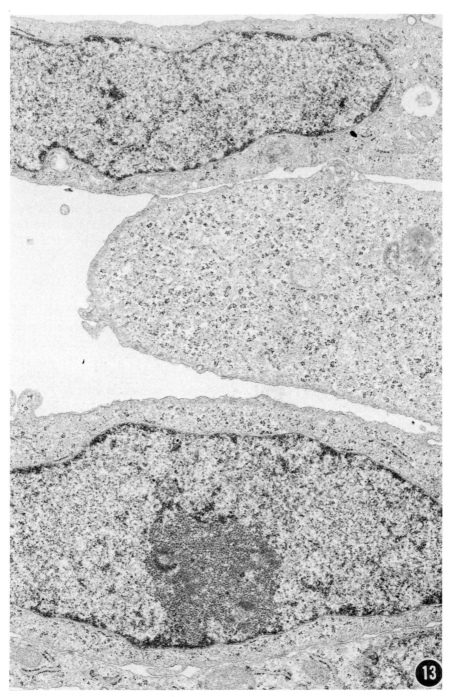

FIG. 13.   Cultured fetal human alveolar fibroblasts (strain HFL-1) in the logarithmic phase of growth contain numerous clusters of free ribosomes; other cytoplasmic organelles are sparse. ×17,000.

human lung produce approximately 4×10^4 collagen chains per hour per cell (Bradley *et al.*, 1975) (Fig. 14). Since fibroblasts are approximately 40% of the total cells present (excluding macrophages and lymphocytes), these data are consistent with the concept that cultured fibroblasts are not "dedifferentiated" with respect to collagen production.

3. Not only do cultured human alveolar fibroblasts produce approximately 6 $\times 10^5$ collagen chains per hour per cell, but they continue to do so for at least 25 population doublings (Breul *et al.*, 1979) (Fig. 14).

If collagen production is taken as a prototypic manifestation of the fibroblast phenotype, it seems reasonable to describe the cultured human alveolar fibroblast as a cell that has sufficient internal controls to maintain its differentiated state *in vitro* and thus as a useful model of its *in vivo* function.

D. Ontogeny of Lung Fibroblasts

In general, the morphology of cells comprising a culture of human alveolar fibroblasts is similar to that of fibroblasts cultured from other soft tissues. This does not prove, however, that they are identical cells. While they may be similar in terms of their major differentiated functions (e.g., production of collagens I and III, collagenase production, fibronectin production), there is increasing evidence that fibroblasts from different organs have subtle differences by which they can be distinguished. For example, lung and skin fibroblasts differ in terms of volume, doubling time, saturation density, life span, and RNA/DNA and protein/DNA ratios (Schneider *et al.*, 1977). Other studies have suggested that these cells also differ in glycosaminoglycan production (Sjöberg and Fransson, 1977) and growth requirements (Ham, this volume).

One way to evaluate the differentiated states of these cells is to examine the

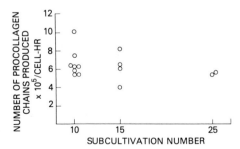

FIG. 14.   Maintenance of the differentiated state of human lung fibroblasts with respect to collagen production. Data derived from cultured human fetal alveolar fibroblasts (strain HFL-1) studied from subcultivations 10–25, a period representing at least 25 population doublings. Normal HFL-1 cells produce approximately 6×10^5 procollagen chains per hour per cell; this differentiated function is well maintained in culture over many population doublings (Breul *et al.*, 1979).

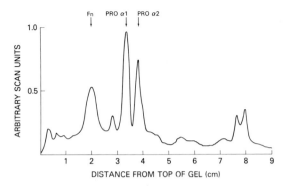

FIG. 15. Proteins secreted by cultured human fetal alveolar fibroblasts (strain HFL-1). Cells were labeled with [^{14}C]proline for 24 hr, and the medium electrophoresed on SDS acrylamide gels where the labeled proteins were separated by molecular weight (Baum *et al.*, 1977; Horwitz *et al.*, 1977). Shown is a scan of the fluorogram of the labeled secreted proteins. The major proteins secreted by these cells are fibronectin (Fn) and procollagens; included in the latter are the pro α1(I), pro α2, and pro α1(III) chains.

array of proteins they secrete into their environment. For human alveolar fibroblasts, although the major secreted proteins are collagens and fibronectin, a number of other proteins, as yet unidentified, are also secreted into the culture medium (Fig. 15). In a comparison of the secreted proteins of human skin and lung fibroblasts, it was apparent that, although the secreted proteins were characteristic for this cell type, there appeared to be an ontological divergence of fibroblasts in terms of this parameter (Fig. 16) (Elson and Crystal, 1977). Most importantly, while human fetal skin and lung fibroblasts generally secreted the same array of proteins, human adult skin and lung fibroblasts had a number of differences in the "minor" (i.e., not collagen or fibronectin) proteins they produced. This kind of observation is consistent with the growing evidence that similar cell types from different organs are distinguishable on biochemical, if not morphological, grounds (Griffin *et al.*, 1976).

E.   Microheterogeneity of Lung Fibroblasts in Culture

Not only is there ontogenic divergence of fibroblasts from different organs, but there is also evidence of microheterogeneity in a population of cultured lung fibroblasts. Within a so-called mass culture of lung fibroblasts are cells with different life spans (Smith and Hayflick, 1974), sizes (Cristofalo and Kritchevsky, 1969), and interdivision times (Absher and Absher, 1976). This is consistent with the suggestion that mass cultures of fibroblasts from human organs such as gingiva may be heterogeneous with respect to this response to environmental stimuli (Narayanan and Page, 1977). However, while such microheterogeneities may exist, current evidence suggests that, on the average, populations of fibro-

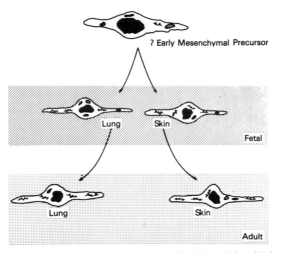

FIG. 16. Current concepts of the ontogeny of human fibroblasts. Most likely derived from an early mesenchymal precursor, fibroblasts of organs such as lung and skin begin to diverge through the various stages of development. Although there are few distinguishing features between fetal fibroblasts from different organs, there are clear biochemical differences in adult lung and skin cells (Elson and Crystal, 1977).

blasts cultured from human alveolar structures are generally similar in terms of their expression of major differentiated functions such as collagen production (Breul *et al.*, 1979). This is consistent with the observations that all fibroblasts cultured from human skin produce collagen and that all produce both collagens I and III (Gay *et al.*, 1976).

F. Identification of Fibroblasts in Culture

Usually morphological criteria are sufficient to distinguish alveolar fibroblasts from other cell populations of the lower respiratory tract. The parallel arrangement of the fibroblasts in culture together with their elongated appearence is quite different from the appearance of epithelial or endothelial cells in culture. Although unlikely, it is conceivable that the smooth muscle cells–pericyte group of mesenchymal cells could predominate in the cultures (particularly if the donor had a lung disorder such as idiopathic pulmonary fibrosis in which islands of smooth muscle cells are characteristic of the lung biopsies) (Crystal *et al.*, 1978). However, the morphology of cultures of smooth muscle cells is clearly distinguishable from that of fibroblasts at the light microscopic level, and definitive at the electron microscopic level.

Another approach in the identification of lung cells in culture is to utilize biochemical markers specific for each cell type. For example, alveolar type II cells produce surfactant, of which dipalmitoylphosphatidylcholine is the major constituent. While type II cells contain major amounts of dipalmitoylphosphatidylcholine (30% of the total phospholipid), alveolar fibroblasts contain one-third the amount (R. Mason, personal communication). Endothelial cells also have specific biochemical markers, the most important of which are the presence of angiotensin-converting enzyme on their surface and the production of factor VIII antigen. It is also possible to distinguish these cells by the general array of proteins they secrete into their environment. While secreted protein "maps" of fibroblasts are specific for these cells, type II cells, macrophages, and other lung cell types clearly secrete a whole array of different proteins (Elson *et al.*, 1976).

G.   Future Studies

To date, most work concerning the alveolar fibroblast as a lung cell has concentrated on its function in maintaining alveolar interstitial connective tissue. It is clear, however, that lung fibroblasts likely play many roles in maintaining alveolar structure and function (Table II). Thus, the increase in the relative number of alveolar fibroblasts in disorders such as oxygen toxicity (Kapanci *et al.*, 1969), or their loss as in destructive disorders such as emphysema, may be a critical determinant of the alterations in lung structure and function in these disorders. It is also apparent that, even though the alveolar fibroblast is the culture prototype of a group of cells found *in vivo,* there may be subtle differences among fibroblasts, interstitial cells, and myofibroblasts that may have profound effects if there were *in vivo* shifts in the relative numbers of each of these cell types. We might expect, therefore, that future culture studies will be able to differentiate among these cell types and to define their relationships and ontogeny. As this information accumulates, an understanding of the contribution of the alveolar fibroblast to the overall structure and function of the lung will emerge.

REFERENCES

Absher, P. M., and Absher, R. G. (1976). *Exp. Cell Res.* **103**, 247–255.
Al-Adnani, M. S., and McGee, J. O.'D. (1976). *Nature (London)* **263**, 145–146.
American Type Culture Collection (1979). Catalogue of Strains II.
Askin, F. B., and Kuhn, C. (1971). *Lab Invest.* **25**, 260–268.
Baenziger, N. L., Dillender, M. J., and Majerus, P. W. (1977). *Biochem. Biophys. Res. Commun.* **78**, 294–301.
Barry, B. E., Crapo, J. D., Gehr, P., Bachofen, M., and Weibel, E. R. (1979). *Am. Rev. Respir. Dis.* **119**, Part 2, 287.
Bartholomew, E. M., Bartholomew, W. R., and Rose, N. R. (1969). *J. Immunol.* **103**, 787–794.
Baum, B. J., McDonald, J. A., and Crystal, R. G. (1977). *Biochem. Biophys. Res. Commun.* **79**, 8–15.

Bernik, M. B., and Kwaan, H. C. (1969). *J. Clin. Invest.* **48,** 1740-1753.

Bloom, S., and Cancilla, P. A. (1969). *Circ. Res.* **24,** 189-196.

Bosmann, H. B., Gutheil, R. L., Jr., and Case, K. R. (1976). *Nature (London)* **261,** 499-501.

Bradley, K., McConnell-Breul, S., and Crystal, R. G. (1975). *J. Clin. Invest.* **55,** 543-550.

Brandes, D., Murphy, D. G., Anton, E. B., and Barnard, S. (1972). *J. Ultrastruct. Res.* **39,** 465-483.

Bray, B. A. (1978). *J. Clin. Invest.* **62,** 745-752.

Breul, S. D., Bradley, K. H., Hance, A. J., Schafer, M. P., Berg, R. A., and Crystal, R. G. (1980). *J. Biol. Chem.* (in press).

Brody, J. S., and Vaccaro, C. (1979). *Fed. Proc. Fed. Am. Soc. Exp. Biol.* **38,** 215-223.

Burri, P. H., and Weibel, E. R. (1977). *In* "Development of the Lung" (W. A. Hodson, ed.), Vol. 6, pp. 215-268. Dekker, New York.

Cristofalo, V. J. (1970). *In* "Aging in Cell and Tissue Culture" (V. J. Cristofalo and E. Holeckovas, eds.), pp. 83-119. Plenum, New York.

Crostofalo, V. J. (1972). *Adv. Gerontol. Res.* **4,** 45-79.

Cristofalo, V. J., and Kritchevsky, D. (1965). *Proc. Soc. Exp. Biol. Med.* **118,** 1109-1113.

Cristofalo, V. J., and Kritchevsky, D. (1966). *J. Cell. Physiol.* **67,** 125-132.

Cristofalo, V. J., and Kritchevsky, D. (1969). *Med. Exp.* **19,** 313-320.

Crystal, R. G., Fulmer, J. D., Baum, B. J., Bernardo, J., Bradley, K. H., Breul, S. D., Elson, N. A., Fells, G. A., Ferrans, V. J., Gadek, J. E., Hunninghake, G. W., Kawanami, O., Kelman, J. A., Line, B. R., McDonald, J. A., McLees, B. D., Roberts, W. C., Rosenberg, D. M., Tolstoshev, P., von Gal, E., and Weinberger, S. E. (1978). *Lung* **155,** 199-224.

Dayer, J. M., Robinson, D. R., and Krane, S. M. (1977). *J. Exp. Med.* **145,** 1399-1404.

Eley, B. M., and Harrison, J. D. (1975). *J. Periodontol. Res.* **10,** 168-170.

Elson, N. A., and Crystal, R. G. (1977). *Clin. Res.* **25,** 356A.

Elson, N. A., Karlinsky, J. B., Kelman, J. A., Rhoades, R. A., and Crystal, R. G. (1976). *Clin. Res.* **24,** 464A.

Franks, L. M., and Cooper, T. W. (1972). *Int. J. Cancer* **9,** 19-29.

Fritz, K. E., Jarmolych, J., and Daould, A. S. (1970). *Exp. Mol. Pathol.* **12,** 354-362.

Fujimoto, W. Y., Teague, J., and Williams, R. H. (1977). *In Vitro* **13,** 237-244.

Fulmer, J. D., and Crystal, R. G. (1979). *In* "Current Pulmonology" (D. H. Simmons, ed.), Vol. 1, pp. 1-65. Houghton, Boston, Massachusetts.

Gay, S., Martin, G. R., Muller, P. K., Timpl, R., and Kuhn, K. (1976). *Proc. Natl. Acad. Sci. U.S.A.* **73,** 4037-4040.

Goldman, R. (1977). *Exp. Cell Res.* **104,** 325-334.

Griffin, J. E., Punyashthiti, K., and Wilson, J. D. (1976). *J. Clin. Invest.* **57,** 1342-1351.

Grove, G., and Cristofalo, V. J. (1976). *J. Cell. Physiol.* **90,** 415-422.

Hance, A. J., and Crystal, R. G. (1975). *Am. Rev. Respir. Dis.* **112,** 657-711.

Hance, A. J., Bradley, K., and Crystal, R. G. (1976). *J. Clin. Invest.* **57,** 102-111.

Hayflick, L. (1965). *Tex. Rep. Biol. Med.* **23,** Suppl. 1, 285-303.

Hayflick, L. (1977). *In* "Handbook of the Biology of Aging" (C. E. Finch and L. Hayflick, eds.), pp. 159-186. Van Nostrand-Reinhold, Princeton, New Jersey.

Hayflick, L., and Moorhead, P. S. (1961). *Exp. Cell Res.* **25,** 585-621.

Hayflick, L., Plotkin, S. A., Norton, T. W., and Koprowski, H. (1962). *Am. J. Hyg.* **75,** 240-258.

Hedman, K., Vaheri, A., and Wartiovaara, J. (1978). *J. Cell Biol.* **76,** 748-760.

Henderson, W. J., Blundell, G., Richards, R., Hext, P. M., Volcani, B. E., and Griffiths, K. (1975). *Environ. Res.* **9,** 173-178.

Holliday, R., and Tarrant, G. M. (1972). *Nature (London)* **238,** 26-30.

Horwitz, A. L., Hance, A. J., and Crystal, R. G. (1977). *Proc. Natl. Acad. Sci. U.S.A.* **74,** 897-901.

Kapanci, Y., and Costabella, P. M. (1979). *Fed. Proc., Fed. Am. Soc. Exp. Biol.* **38**, No. 3, Part 2, 964.

Kapanci, Y., Weibel, E. R., Kaplan, H. P., and Robinson, F. R. (1969). *Lab. Invest.* **20**, 101-118.

Kapanci, Y., Assimacopoulos, A., Irle, C., Zwahlen, A., and Gabbiani, G. (1974). *J. Cell Biol.* **60**, 375-392.

Kapanci, Y., Costabella, P. M., and Gabbiani, G. (1976). *In* "Lung Cells in Disease" (A. Bouhuys, ed.), pp. 69-84. North-Holland Publ., Amsterdam.

Kelley, J., Beaumont, R. T., and Moehring, J. M. (1979). *Am. Rev. Respir. Dis.* **119**, No. 4, Part 2, 323.

Kelman, J., Brin, S., Horwitz, A., Bradley, K., Hance A., Breul, S., Baum, B., and Crystal, R. (1977). *Am. Rev. Respir. Dis.* **115**, Part 2, 343.

Kronenberg, L. H. (1977). *Virology* **76**, 634-642.

Kuhn, C. (1976). *In* "The Biochemical Basis of Pulmonary Function" (R. G. Crystal, ed.), pp. 3-48. Dekker, New York.

Laemmli, U. K. (1970). *Nature (London)* **227**, 680-685.

Lane, B. P. (1965). *J. Cell Biol.* **27**, 199-213.

Laskey, R. A., and Mills, A. D. (1975). *Eur. J. Biochem.* **56**, 335-341.

Lazarides, E. (1976). *In* "Cold Spring Harbor Conference on Cell Proliferation" (R. Goldman, T. Pollard, and J. Rosenbaum, eds.), Vol. 1, pp. 347-360. Cold Spring Harbor Lab., Cold Spring Harbor, New York.

McDonald, J. A., Baum, B. J., Rosenberg, D. M., Kelman, J. A., Brin, S. C., and Crystal, R. G. (1979). *Lab. Invest.* **40**, 350-357.

Macieira-Coelho, A., Pontén, J., and Philipson, L. (1966a). *Exp. Cell Res.* **42**, 673-684.

Macieira-Coelho, A., Pontén, J., and Philipson, L. (1966b). *Exp. Cell Res.* **43**, 20-29.

Madri, J. A., and Furthmayr, H. (1979). *Am. J. Pathol.* **94**, 323-332.

Mathé, A., Hedqvist, P., Strandberg, K., and Leslie, C. A. (1977). *N. Engl. J. Med.* **296**, 850-855 and 910-914.

Min, K.-W. (1979). *Fed. Proc., Fed. Am. Soc. Exp. Biol.* **38**, No. 3, Part 2, 923.

Mosher, D. F., and Wing, D. A. (1976). *J. Exp. Med.* **143**, 462-467.

Mosher, D. F., Saksela, O., and Vaheri, A. (1977). *J. Clin. Invest.* **60**, 1036-1045.

Mott, D. M., Fabisch, P. H., Sani, B. P., and Sorof, S. (1974). *Biochem. Biophys. Res. Commun.* **61**, 621-627.

Narayanan, A. S., and Page, R. C. (1977). *FEBS Lett.* **80**, 221-224.

O'Hare, K. H., Reiss, O., and Vatter, A. (1971). *J. Histochem. Cytochem.* **19**, 97-115.

Pearlstein, E., Hynes, R. O., Franks, L. M., and Hemmings, V. J. (1976). *Cancer Res.* **36**, 1475-1480.

Reid, K. B. M., and Solomon, E. (1977). *Biochem. J.* **167**, 647-660.

Reynolds, H. Y., Fulmer, J. D., Kazmierowski, J. A., Roberts, W. C., Frank, M. M., and Crystal, R. G. (1977). *J. Clin. Invest.* **59**, 165-175.

Rhodin, J. A. G. (1968). *J. Ultrastruct. Res.* **25**, 452-500.

Rose, G. G., Yajima, T., and Mahan, C. J. (1978). *J. Dent. Res.* **57**, 1003-1015.

Ross, R. (1968a). *In* "Treatise on Collagen" (B. S. Gould, ed.), Vol. 2, Part A, pp. 2-82. Academic Press, New York.

Ross, R. (1968b). *Biol. Rev. Cambridge Philos. Soc.* **43**, 51-96.

Rothblat, G. H., Boyd, R., and Deal, C. (1971). *Exp. Cell Res.* **67**, 436-440.

Ryan, G. B., Cliff, W. J., Gabbiani, G., Irle, C., Stratkov, P. R., and Majno, G. (1973). *Lab. Invest.* **29**, 197-206.

Schneider, E. L., and Shorr, S. S. (1975). *Cell* **6**, 179-184.

Schneider, E. L., Mitsui, Y., Au, K. S., and Shorr, S. S. (1977). *Exp. Cell Res.* **108**, 1-6.

Schwarz, R. I., and Bissell, M. J. (1977). *Proc. Natl. Acad. Sci. U.S.A.* **74**, 4453-4457.

such as membrane filters on stainless steel grids (Resnick *et al.*, 1974), lens paper (Davis, 1967), watch glasses (Kolesnichenko and Nikonova, 1975), gelatin sponge (Ekelund *et al.*, 1975; Stoner *et al.*, 1978), and collagen-coated glass coverslips (Rose and Yajima, 1977). In the majority of these studies, the normal architecture of the tissue was lost after a few days of incubation, and the central areas of the tissue rapidly became necrotic, presumably because of the lack of oxygen and nutrients and the accumulation of toxic products. Nevertheless, short-term maintenance of peripheral lung in organ culture has permitted a wide variety of investigations such as analyses of the effects of varying concentrations of oxygen on the integrity of the tissue (Resnick *et al.*, 1974), of the incorporation of precursors into surfactant phospholipids (Gross *et al.*, 1978), of the ability of the tissue to metabolize the environmental procarcinogen benzo[*a*]pyrene (Stoner *et al.*, 1978), and of the effects of marijuana and tobacco smoke on DNA synthesis and chromosomal complement in lung cells (Leuchtenberger and Leuchtenberger, 1970; Leuchtenberger *et al.*, 1973).

In the present chapter, a technique used in our laboratory for the short-term culture of explants of peripheral lung from adult humans is described. Fragments of lung were placed on pieces of gelatin sponge in tissue culture dishes, and the cultures put in a rocking chamber to permit intermittent exposure of the tissue to the atmosphere and to the culture medium. With this method, the normal architecture of the lung was maintained for 3–5 days, and the metabolism of benzo[*a*]pyrene by the various lung cell types was determined by both biochemical methods and autoradiography (Stoner *et al.*, 1978).

II.   Materials and Methods

A.   Preparation of Culture Dishes Containing Gelatin Sponge

Gelatin sponge (Gelfoam,[1] 12 cm^2, 3 mm thick; The Upjohn Company, Kalamazoo, Michigan) was cut into disks 1 cm in diameter. The disks were stored at 4°C in L-15 (Leibovitz, 1963) tissue culture medium (Grand Island Biological Company, Grand Island, New York). At the time of culture, two gelatin sponge disks were placed opposite each other in 60 × 15 mm tissue culture dishes (Falcon Plastics Company, Oxnard, California) and on areas of the dish that had previously been etched with a sterile needle. The dishes were then placed in an incubator at 36.5°C for 1 hour to warm the gelatin.

B.   Lung Explant Cultures

Nontumorous peripheral lung specimens were obtained at the time of either surgery or "immediate" autopsy (Trump *et al.*, 1974). The specimens were

[1]Registered Trademark.

immediately immersed in L-15 medium at 4°C for transport to the laboratory. The interval between obtainment of specimens and transport to the laboratory was approximately 4 hours. Tissue from the outer edges of the lung was cut into pieces of 2-3 mm³, using two scalpels in scissor-like fashion; 50–60 explants were obtained from each case. Two or three pieces were placed on each gelatin disk, and 3 ml of CMRL-1066 tissue culture medium (Grand Island Biological Company) supplemented with 5% heat-inactivated fetal bovine serum, $1.0\,\mu g/ml$ of bovine insulin (Eli Lilly and Company, Indianapolis, Indiana), $0.1\,\mu g/ml$ of hydrocortisone (The Upjohn Company), $0.1\,\mu g/ml$ of β-retinyl acetate (Hoffman-LaRoche, Inc., Nutley, New Jersey), and $50\,\mu g/ml$ of gentamicin was added to each dish. The dishes were placed in a controlled-atmosphere chamber (Bellco Glass Company, Vineland, New Jersey), as described by Barrett et al. (1976), and gassed at 10 psi with a mixture of 50% oxygen–45% nitrogen–5% carbon dioxide for approximately 5 minutes. The chamber was then placed on a rocker platform (Bellco Glass Company) which rocked at 10 cycles per minute, causing the medium to flow intermittently over the surface of the explant. The explants were incubated at 36.5°C. The medium and atmosphere were changed every 2–3 days.

C. Light and Electron Microscopy

At daily intervals, four or five explants were fixed in a phosphate-buffered solution of 4% formaldehyde and 1% glutaraldehyde (4FIG), pH 7.4 (McDowell and Trump, 1976), postfixed in cold 1.33% osmium tetroxide buffered with s-collidine (pH 7.4) for 2 hours, dehydrated in a graded series of ethanol, and embedded in Epon–Araldite (Harris et al., 1972). Sections (1 μm) were cut and stained with toluidine blue. Areas of interest in thick sections were chosen for thin-sectioning. Thin sections (600Å) were double-stained on the grids with lead acetate and uranyl nitrate and examined in a Hitachi HU-12A electron microscope.

D. Autoradiography

At various time intervals explants were incubated for 2 hours with [³H]thymidine ($25\,\mu Ci/ml$; specific activity, 20 Ci/mmole). After both fixation in 4FIG and embedding in Epon–Araldite, 1-μm sections were coated with NTB-2 photographic emulsion (Eastman Kodak Company, Rochester, New York) at 45°C, dried, and stored in the dark at 4°C with dessicant for 3 weeks. The autoradiograms were developed and fixed as previously described (Boren et al., 1974).

The labeling index (percentage of cells with labeled nuclei) was obtained by counting the number of both alveolar lining cells and bronchiolar cells in three explants (nine plastic sections) per time interval and recording the total number and the number labeled. Nuclei overlayed with 20 or more grains were considered labeled. Generally, from 3000 to 5000 cells of each type were counted at each time interval.

III. Results

A. Light Microscopy

Prior to culture, the lung specimens consisted of typical alveolar structures lined with epithelial cells (Fig. 1). The alveoli in specimens from some patients were nearly filled with macrophages, whereas in other cases they contained only a small number of macrophages. Connective tissue composed of fibroblast cells, collagen, and elastin often surrounded the alveolar structures. Terminal bronchiolar cells and endothelial cells were also observed.

Human lung tissues maintained for at least 3 days in explant culture were similar in appearance to the tissue immediately prior to culture. In explants 2–3 mm^3 in size, the alveolar walls remained distinct throughout the entire tissue. In explants of larger size, i.e., 4–6 mm^3, the structural integrity of the alveoli was lost in the central portions of the tissue as a result of collapse of the alveolar walls. Therefore, it was necessary to culture smaller pieces of tissue to maintain the structural integrity of the alveoli throughout the explant. There was also a marked thickening of the alveolar walls during the first few days in culture presumably resulting from an increase in both the size and number of epithelial lining cells (Fig. 2). Bronchiolar epithelial cells and endothelial cells were well preserved during this period.

After 3 days, the histological appearance of the cultured explants was variable. The central portions of most explants eventually became necrotic; however, necrosis was rarely observed in the peripheral portions of the explant, i.e., those areas either exposed to the atmosphere or overlying the gelatin sponge. During the period of culture, in the majority of explants, there was a progressive migration of epithelial cells around the outer edge of the explant (Fig. 3). Most of these cells contained cytoplasmic inclusions that stained darkly with toluidine blue. With increasing time of incubation, there was a progressive tendency for the alveolar structures to collapse and form a solid mass of tissue. However, in the peripheral portions of the explant, intact alveolar structures could be observed in tissue cultured for as long as 25 days.

After 7–10 days, the outer portion of most explants was composed of intact alveolar structures lined with cells that contained cytoplasmic inclusions which stained darkly with toluidine blue (Fig. 4). Some cells had numerous inclusions, others only a few, and inclusions were observed in the alveolar spaces. In the internal portions of the explant where the alveolar structures had collapsed, the cells with inclusions were necrotic and adjacent to one another. The inclusions of

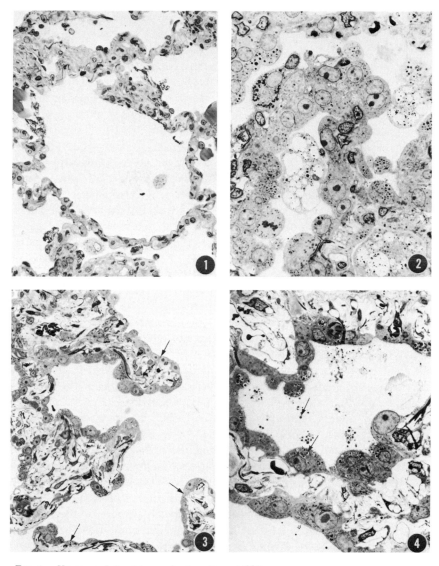

FIG. 1.   Human peripheral lung prior to culture. ×330.

FIG. 2.   Human lung tissue, day 3 of culture. Note thickening of alveolar lining and degeneration of macrophages. ×880.

FIG. 3.   Human lung tissue, day 7 of culture. Note migration of cells (arrows) around outer edges of explant. ×330.

FIG. 4.   Human lung tissue, day 7 of culture. Numerous cytoplasmic inclusions (arrows) are in the alveolar lining cells and in the alveolar space. ×880.

these cells were usually much larger than those observed in cells lining the intact alveolar structures in the outer portions of the explant (Fig. 5). Normal-appearing fibroblasts proliferated in the internal portions of the explant, and these areas progressively became filled with large amounts of connective tissue composed of collagen and elastin. Alveolar epithelial cells with inclusions, bronchiolar epithelial cells, and connective tissue cells were the predominant cell types in explants maintained for 25 days. Epithelial cells with inclusions were more prevalent in the peripheral portions of the explants, and connective tissue cells in the internal portions. No attempt was made to culture the explants for longer than 25 days.

Alveolar macrophages generally became necrotic after 3–7 days in culture. Explants that contained a large number of macrophages prior to culture did not appear to survive as well as those that contained few macrophages.

Cells of all types, including epithelial cells with inclusions, macrophages, bronchiolar cells, and fibroblasts migrated into the upper matrix of the gelatin sponge during culture (Fig. 6). The cells in the sponge were well maintained during 25 days *in vitro*.

B.   Electron Microscopy

The alveoli of fresh tissue contained type-1 epithelial cells with extended cytoplasm and type-2 epithelial cells. Gaseous exchange occurs across the cytoplasm of the type-1 cells. Type-2 cells contained intracytoplasmic lamellar inclusion bodies that corresponded to the darkly stained inclusions observed in the cytoplasm of these cells with light microscopy. Lamellar bodies are composed, in part, of tubular myelin and are the intracellular site of the storage of pulmonary surfactant (Adamson and Bowden, 1973; Buckingham *et al.*, 1966; Chevalier and Collet, 1972; Kikkawa *et al.*, 1965). Surfactant, a complex mixture of phospholipids, carbohydrates, proteins, and nucleic acids, lines the alveolar walls and reduces the surface tension across the alveolar interface, thus preventing alveolar collapse at low lung volumes. Type-2 cells also contained endoplasmic reticulum, mitochondria, Golgi complex, and numerous microvilli. The alveoli of fresh lung also contained variable numbers of macrophages and were surrounded by interstitial tissue composed of fibroblasts, collagen, elastin, and reticular cells. Tissues from some cases contained large amounts of collagen, suggesting the development of pulmonary fibrosis.

Explants cultured for 5–25 days contained alveolar structures composed largely of type-2 epithelial cells directly adjacent to one another (Fig. 7). It appeared that the type-2 epithelial cells proliferated during the period of culture and gradually replaced the type-1 cells. This phenomenon of the close association of type-2 cells occurred in the peripheral portions of the explant with intact alveolar structures and in the internal portions of the explant where the alveoli had collapsed. Type-2 cells in the peripheral portions of the explant contained

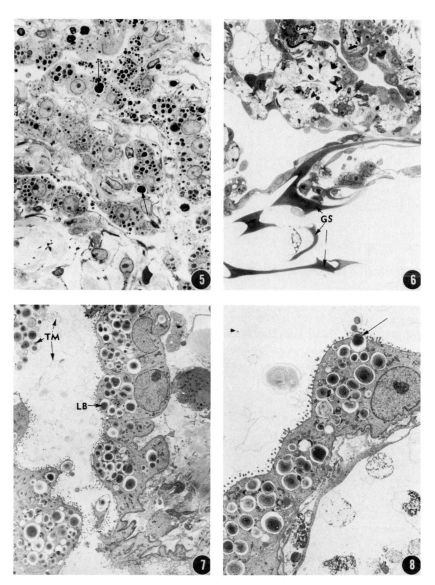

FIG. 5. Human lung tissue, day 17 of culture. Alveolar structures in this portion of explant have collapsed. Numerous large cytoplasmic inclusions (arrows) are in epithelial cells. ×880.

FIG. 6. Human lung tissue, day 7 of culture. Cells are beginning to migrate from the explant into the gelatin sponge (GS). ×330.

FIG. 7. Electron micrograph of human alveolar type-2 epithelial cells in explant cultured for 10 days. LB, Lamellar bodies; TM, tubular myelin in alveolar space. ×3500.

FIG. 8. Electron micrograph illustrating the early stage of the release of a lamellar inclusion body from a type-2 cell. ×7200.

lamellar inclusion bodies that were generally rounded, variable in size, and occasionally observed in the process of being extruded from the cell into the alveolar space (Fig. 8). In compact areas of explants where alveoli had collapsed, fewer lamellar bodies were released from the type-2 cells and, in fact, they accumulated in the cytoplasm and often became enlarged and/or coalesced with other lamellar bodies. As mentioned above, cells grew around the outer edge of a majority of the explants during the period of culture. Electron microscopy revealed that most of these cells were type-2 cells. Fibroblast cells proliferated in the internal portions of the explants during culture and produced large amounts of collagen which was frequently seen in the cytoplasm (Figs. 9 and 10). Bronchiolar epithelial cells also proliferated during culture and were well preserved in the peripheral portions of the explant for at least 25 days (Fig. 11).

C.   Autoradiography

The labeling index of alveolar lining cells and bronchiolar epithelial cells after various periods in culture is given in Table I. Generally, 1–2% of the alveolar or bronchiolar epithelial cells were labeled after the various periods of incubation (Fig. 12). The majority of the alveolar lining cells contained darkly stained inclusions identified as lamellar bodies by electron microscopy. There were also labeled alveolar lining cells without inclusions. These were probably undifferentiated type-2 cells, since type-1 cells do not undergo cell division. The incorporation of [^{3}H]thymidine into the nuclei of type-2 cells in the peripheral portions of the explants occurred with much greater frequency than into the nuclei of type-2 cells in the internal portions where the alveolar structures had collapsed.

Lung fibroblasts were also labeled after exposure to [^{3}H]thymidine, however, the labeling index for these cells was not determined. Labeled nuclei were rarely seen in endothelial cells and alveolar macrophages.

IV.   Discussion

The method of explant culture used in our laboratory for the long-term cultivation of human bronchus (Barrett et al., 1976), pancreatic duct (Jones et al., 1977), and colon (Autrup et al., 1978) has been applied, with two modifications, in the culture of peripheral lung. One modification was the placing of the tissues on a gelatin sponge, which permitted their survival adjacent to the sponge and the migration of cells into the sponge to establish sponge cultures. This procedure was undertaken after preliminary experiments indicated that culture of the tissues on plastic led to rapid deterioration of the cells adjacent to the plastic. Undoubtedly, the sponge matrix permitted adequate diffusion of nutrients and oxygen to

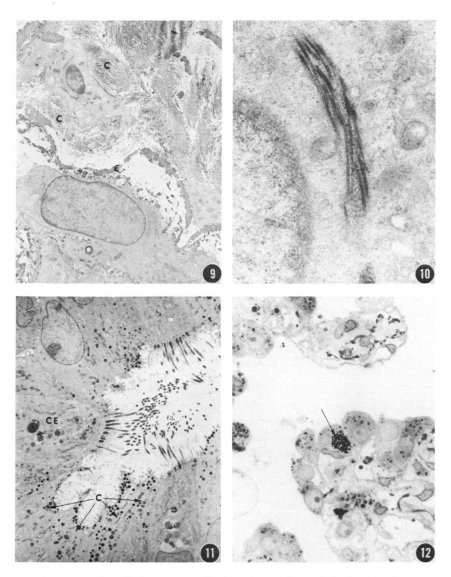

FIG. 9.   Human lung fibroblasts surrounded by collagen (C) and elastin (E). ×2400.
FIG. 10.   Human lung fibroblast, showing intracellular collagen. ×21,000.
FIG. 11.   Human bronchiolar epithelial cells, day 17 of culture. Note ciliated epithelial cells (CE) and Clara cells (C). ×6300.
FIG. 12.   Labeled nuclei in alveolar epithelial cell with darkly stained inclusion bodies. ×880.

TABLE I

Labeling Index of Alveolar Lining Cells and Bronchiolar Cells during Incubation of Human Lung Explants[a]

	[^{3}H]thymidine-labeled (%)	
Incubation time	Alveolar lining cells	Bronchiolar cells
2 hours	0	0
2 days	1.3 ± 0.18	1.0 ± 0.04
4 days	1.1 ± 0.15	2.1 ± 0.10
7 days	1.3 ± 0.10	2.6 ± 0.25
10 days	1.0 ± 0.07	1.1 ± 0.15
17 days	2.3 ± 0.54	1.1 ± 0.20

[a] Values are means plus or minus standard error.

the cells of the explant. The other modification was the culture of explants smaller (2–3 mm^3) than those used for the bronchus and colon (1 cm^2). This modification was necessary because of the tendency of the central areas of larger explants rapidly to become necrotic, presumably because of inadequate oxygenation and diffusion of nutrients and because of the accumulation of toxic metabolites. Even with these modifications, there was necrosis in the internal portions of the explants after a few days in culture. Apparently, this problem did not occur in the studies of Guerrero et al. (1977), who infused agar into the entire lung of the rat prior to explant culture, and of Rose and Yajima (1977), who reported long-term survival of fetal mouse lung using a dual rotary circumfusion system. In the circumfusion system described by Rose and Yajima (1977), the histotypic structure of the terminal bronchiolar and alveolar epithelium was well maintained for a period of at least 75 days. To the author's knowledge, this is the longest period in which the normal appearance of peripheral lung tissue has been maintained in organ culture.

In the present study, the alveolar walls remained intact in the peripheral portions of the explants for relatively long periods of time. Even in areas where the alveolar structure was lost, most of the cells survived for a period of 7–10 days. In addition, the ability of the human lung explants to metabolize benzo[a]pyrene remained quite constant during 7 days of cultivation (Stoner et al., 1978). Similar results on the survival of lung tissues were obtained by Kolesnichenko and Nikonova (1975), who cultured explants of human embryonic lung for 30 days by a modified watch glass method, and by Davis (1967), who maintained explants of guinea pig lung on lens paper for 14 days.

Autoradiographic studies with [^{3}H]thymidine indicated that alveolar lining cells, bronchiolar epithelial cells, and fibroblasts synthesized DNA during at least 17 days in culture. Data are not available for periods of longer than 17 days.

Under the light microscope, the labeled alveolar lining cells could be classified either as (1) cells that contained cytoplasmic inclusions that stained darkly with toluidine blue (type-2 cells), or (2) cells that did not contain stainable inclusions. Since, by electron microscopy, most of the alveolar lining cells in explants maintained for at least 7 days were type-2 cells, the labeled cells without inclusions were probably undifferentiated type-2 cells. Labeled nuclei were observed predominantly in type-2 cells either lining intact alveolar structures or migrating around the edge of the explant. Type-2 cells in areas of the explant where the alveolar structures had collapsed were not labeled. As suggested by Kauffman (1976), these cells had probably degenerated. The labeling index of bronchiolar epithelial cells was significantly higher ($p = <0.05$) than that of the alveolar epithelial cells at 4 and 7 days, and significantly lower ($p = <0.05$) at 17 days. Although the labeling index of the fibroblast cells was not determined, they incorporated thymidine throughout the duration of culture.

Morphological evidence that the type-2 alveolar cells were producing lung surfactant was as follows: (1) There was an increase in both the size and number of lamellar inclusion bodies in type-2 cells during the culture period. As indicated earlier, lamellar bodies appear to be the cellular site for the synthesis and/or storage of lung surfactant. (2) Lamellar bodies were released into the alveolar spaces during culture. Therefore, short-term culture of explants of peripheral lung, by the method described in this chapter, should permit the biochemical analysis of surfactant synthesis. Indeed, similar methods of explant culture have already been employed to examine the incorporation of radiolabeled precursors into surfactant phospholipids (Ekelund et al., 1975; Gross et al., 1978).

In a preliminary study, we found that a gaseous mixture containing 95% oxygen–5% carbon dioxide was very toxic for the human lung explants. Similarly, Resnick et al. (1974) found that a high oxygen environment was very toxic for fetal mouse lung in organ culture. These observations differ from that of Gross et al. (1978) who reported that an atmosphere of 95% oxygen–5% carbon dioxide was not toxic for fetal rat lung in organ culture. It is possible that the inconsistency in these results is due to the methods used for culture of the tissues and/or to species differences in oxygen requirements.

In the present study, it is not known whether insulin, hydrocortisone, or β-retinyl acetate influenced the survival and growth of the lung cells in the explants. Moreover, it was not determined whether the concentrations of these supplements were optimal for preservation of the explants. Further studies are required to determine the answers to these questions.

The survival of human lung tissues in explant culture was influenced by the period of time between collection of the tissue and its placement in culture. The best results were obtained when the tissue was cultured within 6 hours after collection. Tissues kept for as long as 24 hours in transport medium usually became necrotic and did not survive well in culture.

V   Perspectives

The lung is a complex organ estimated to contain at least 40 cell types (Sorokin, 1970). Because of this complexity, it has been difficult to determine the contribution of specific cell types to lung physiology when conducting biochemical studies using lung homogenates. Therefore, during the last several years, considerable effort has been expended toward the development of methods for isolating and culturing specific lung cell types for a variety of investigations. Relatively fewer studies have been concerned with improving methods for the organ or explant culture of lung tissues in spite of the fact that these methods, when compared to monolayer cell culture, have the advantage of maintaining normal cell-to-cell interactions which are important in physiological processes. In addition, techniques such as autoradiography, immunofluorescence, immunoperoxidase and immuno electron microscopy make it possible to assign specific functions to individual cell types. Therefore, organ and explant cultures, as well as histotypic and monolayer cell cultures, can be powerful tools for examining lung cell functions.

The method of explant culture described in this chapter permits the short-term culture of peripheral lung and has been used for a study on the metabolism of benzo[a]pyrene (Stoner et al., 1978). It is likely that the method would also be useful for studies on the biosynthesis of surfactant and collagen, and on the effects of various agents on the synthesis of these products. Because of the loss of cell viability in the internal areas of the explants after long periods in culture, this method would not be useful for long-term studies. In this regard, it would be interesting to determine whether the circumfusion system of Rose and Yajima (1977) would permit longer term maintenance of human lung explants.

ACKNOWLEDGEMENTS

The author thanks Drs. Benjamin Trump and Paul Schafer for supplying the lung tissue, and Dr. Curtis Harris for suggestions in the conduct of the studies. The technical assistance of Frank Jackson, Gwendolyn Myers, and Maria Yamaguchi and the secretarial assistance of Sandy Dorfman are gratefully acknowledged. Permission to reproduce the photographs in Figures 1–7 from the publication entitled, "Explant Culture of Human Peripheral Lung. I. Metabolism of Benzo(a)pyrene" by Stoner et al. (1978) was granted by the U.S.–Canadian Division of the International Academy of Pathology.

REFERENCES

Adamson, I. Y. R., and Bowden, D. H. (1973). Exp. Mol. Pathol. 18, 112–124.
Autrup, H., Barrett, L. A., Jackson, F. E., Jesudason, M. L., Stoner, G., Phelps, P., Trump, B. F., and Harris, C. C. (1978). Gastroenterology 74, 1248–1257.

Barrett, L. A., McDowell, E. M., Frank, A. L., Harris, C. C., and Trump, B. F. (1976). *Cancer Res.* **36**, 1003–1010.

Boren, H., Wright, E., and Harris, C. (1974). *In* "Methods in Cell Biology" (D. Prescott, ed.), Vol. 8, pp. 277–288. Academic Press, New York.

Buckingham, S., Heinemann, H. O., Sommers, S. C., and McNary, W. F. (1966). *Am. J. Pathol.* **48**, 1027–1041.

Chevalier, G., and Collet, A. J. (1972). *Anat. Rec.* **174**, 289–310.

Davis, J. M. G. (1967). *Br. J. Exp. Pathol.* **48**, 371–378.

Ekelund, L., Arvidson, G., Emanuelsson, H., Myhrberg, H., and Astedt, B. (1975). *Cell Tissue Res.* **163**, 263–272.

Gross, I., Walker Smith, G. J., Maniscalco, W. M., Czajka, M. R., Wilson, C. M., and Rooney, S. A. (1978). *J. Appl. Physiol.* **45**, 355–362.

Guerrero, R. R., Rounds, D. E., and Booher, J. (1977). *In Vitro* **13**, 517–524.

Harris, C. C., Sporn, M. B., Kaufmann, D. G., Smith, J. M., Jackson, F. E., and Saffiotti, U. (1972). *J. Natl. Cancer Inst.* **48**, 743–761.

Jones, R. T., Barrett, L. A., van Haaften, C., Harris, C. C., and Trump, B. F. (1977). *J. Natl. Cancer Inst.* **58**, 557–565.

Kauffman, S. L. (1976). *Cell Tissue Kinet.* **9**, 489–497.

Kikkawa, Y., Motoyama, E. K., and Cook, C. D. (1965). *Am. J. Pathol.* **47**, 877–903.

Kolesnichenko, T. S., and Nikonova, T. V. (1975). *Byull. Eksp. Biol.* **80**, 99–102.

Leibovitz, A. (1963). *Am. J. Hyg.* **78**, 173–180.

Leuchtenberger, C., and Leuchtenberger, R. (1970). *Exp. Cell Res.* **62**, 161–172.

Leuchtenberger, C., Leuchtenberger, R., and Schneider, A. (1973). *Nature (London)* **241**, 137–139.

McDowell, E. M., and Trump, B. F. (1976). *Arch. Pathol. Lab. Med.* **100**, 405–414.

Resnick, J. S., Brown, D. M., and Vernier, R. L. (1974). *Lab. Invest.* **31**, 665–677.

Rose, G. G., and Yajima, T. (1977). *In Vitro* **13**, 749–768.

Sorokin, S. P. (1970). *In* "Morphology of Experimental and Respiratory Carcinogenesis" (P. Nettesheim, Hanna, M. G., Jr., Deatherage, J. W., Jr., eds.), pp. 3–41. U.S. At Energy Comm., Washington, D.C.

Stoner, G. D., Harris, C. C., Autrup, H., Trump, B. F., Kingsbury, E. W., and Myers, G. A. (1978). *Lab. Invest.* **38**, 685–692.

Trump, B. F., Valigorsky, J. M., Dees, J. H., Mergner, W. J., Kim, K. M., Jones, R. T., Pendergrass, R. E., Garbens, J., and Cowley, R. A. (1974). *Hum. Pathol.* **4**, 89–109.

Chapter 5

Maintenance of Human and Rat Pulmonary Type II Cells in an Organotypic Culture System [1]

WILLIAM H. J. DOUGLAS, RONALD L. SANDERS, AND KAREN R. HITCHCOCK

Department of Anatomy,
Tufts University School of Medicine,
Boston, Massachusetts

I. Introduction

The lungs of mammals and some inframammalian vertebrates contain a potent surface-active material that coats the alveolar surface, decreases the surface

[1]This investigation was supported in part by Public Health Service grants HL 21008-02, HL 23969-01, and HL 15316.

tension, and thus stabilizes the alveoli against collapse. The presence of this pulmonary surfactant is essential for normal pulmonary function, and its absence or diminution is an important etiological factor in respiratory distress syndrome (King, 1974). Biochemical analyses indicate that surfactant is a complex mixture of phospholipids, neutral lipids, proteins and, possibly, carbohydrates; the major phospholipid present is dipalmitoylphosphatidylcholine (DPPC) (Frosolono *et al.*, 1970; King, 1974; King and Clements, 1972; Van Golde, 1978). In contrast to whole lung tissue, isolated surfactant contains a higher percentage of phospholipids which includes substantial amounts of phosphatidylcholine (PC) and phosphatidylglycerol (PG) (King and Clements, 1972; Sanders and Longmore, 1975; Van Golde, 1978). In addition, a high percentage of the PC molecules contain saturated fatty acids at both the α and β positions (King, 1974; King and Clements, 1972), a relatively uncommon occurrence for PCs.

The mammalian pulmonary alveolus is lined with an endodermally derived epithelium consisting of two cell types, type I and type II pulmonary epithelial cells (Meyrick and Reid, 1977). It has been established that type II cells are secretory and the source of pulmonary surfactant (Mason *et al.*, 1977a). Type II cells are characterized by cytoplasmic inclusions referred to as lamellar bodies. These structures have been implicated in surfactant metabolism, perhaps as sites of surfactant storage (Adamson and Bowden, 1973; Askin and Kuhn, 1971; Massaro and Massaro, 1972). Electron microscopic investigations reveal lamellar-like arrays within the alveolar space, which resemble the lamellar bodies within the type II cells. Extracellular conversion of these secreted lamellar bodies into the tubular myelin found in the surfactant lining layer has been suggested (Williams, 1977).

Possible precursors [i.e., multivesicular bodies; see Meyrick and Reid, (1977)] of mature lamellar bodies are located initially in the basal regions of differentiating epithelial cells; more mature lamellar bodies are found in the apical cytoplasm of type II cells later in gestation. The appearance of lamellar bodies within the future alveolar spaces never precedes the occurrence of these structures within type II cells. Histochemical studies indicate that lamellar bodies, like surfactant, contain lipid and phospholipid (Dermer, 1970; Sorokin, 1961), acid mucopolysaccharides (Kalifat *et al.*, 1970), and surfactant apoprotein (King, 1977). Morphological and chemical analyses of purified fractions of lamellar bodies have further documented the probable relationship of type II cell lamellar bodies and the surfactant system of the lung (Gil and Reiss, 1973; Hallman *et al.*, 1976). Indeed, intact isolated lamellar bodies have been shown to form surface films with surface properties similar to those of surfactant (Paul *et al.*, 1977).

Earlier studies of surfactant phospholipid synthesis generally employed *in vivo* methods, *ex vivo* perfusion systems, slices, explants, or mixed lung cell cultures. With these techniques, a number of investigators found that a variety of precursors could be used for pulmonary phospholipid synthesis, including glucose (Felts, 1964; Frosolono *et al.*, 1976; Godinez *et al.*, 1975; Sanders and Longmore, 1975), glycerol (Mims *et al.*, 1975), acetate (Felts, 1964; Godinez *et al.*, 1975; Sanders and Longmore, 1975), lactate (Frosolono *et al.*, 1976; Godinez *et al.*, 1975), palmitate (Felts, 1964; Sanders and Longmore, 1975), and choline (Funkhouser and Hughes, 1977; Gross *et al.*, 1978; Smith and Torday, 1974). However, because the lung is composed of approximately 40 cell types (Sorokin, 1970), the precise mechanisms of surfactant synthesis in type II cells cannot be obtained from studies that utilize systems in which the majority of cells are not involved directly in surfactant synthesis.

Some recent investigations of lung lipid synthesis have used culture systems of isolated cells with morphological characteristics of type II pneumocytes (Batenburg *et al.*, 1978; Douglas *et al.*, 1976a; Douglas and Teel, 1976; Frosolono *et al.*, 1976; Smith and Kikkawa, 1978; Spitzer *et al.*, 1975; Wykle *et al.*, 1977). These cell preparations produce PC from a number of precursors, including acetate (Batenburg *et al.*, 1978; Smith and Kikkawa, 1978; Wykle *et al.*, 1977), glycerol (Batenburg *et al.*, 1978; Smith and Kikkawa, 1978; Wykle *et al.*, 1977), choline (Batenburg *et al.*, 1978; Frosolono *et al.*, 1976; Smith and Kikkawa, 1978; Spitzer *et al.*, 1975), palmitate (Batenburg *et al.*, 1978; Frosolono *et al.*, 1976; Smith and Kikkawa, 1978; Wykle *et al.*, 1977), and glucose (Batenburg *et al.*, 1978; Mason *et al.*, 1977a,b; Smith and Kikkawa, 1978). Isolated cells are also capable of disaturated PC (DSPC) synthesis (Batenburg *et al.*, 1978; Smith and Kikkawa, 1978; Spitzer *et al.*, 1975; Wykle *et al.*, 1977). As in previous studies, the predominant route of PC synthesis is the CDP-choline pathway (Douglas *et al.*, 1976a). Although these studies have advanced our knowledge of lipid metabolism in type II cells, interpretation of their data is complicated by the fact that only total cell lipids or DSPC was analyzed. If the mechanisms involved in the synthesis of surfactant are to be elucidated, surfactant phospholipids must be separated from other cellular components. Use of these cell systems is limited in that (1) not all preparations are homogeneous for a single cell type, (2) they are established from adult lung and do not permit studies of the initiation of surfactant synthesis, and (3) they are only suitable for short-term studies of 12–48 hours.

Culture techniques developed by Moscona and Moscona (1952; Moscona, 1962) permit dissociation of fetal tissue into single, viable cells that retain the capacity to reaggregate and form histotypic structures *in vivo*. This capacity of histotypic reaggregation is characteristic of numerous fetal tissues including liver (Sankaran *et al.*, 1977), heart (Fischman and Moscona, 1971), and brain (Garber and Moscona, 1972).

Ansevin and Lipps (1973) utilized these techniques to obtain histotypic cultures of chick embryo kidney. The kidney cells were capable of reaggregation, and histotypic structures formed that retained renal-specific functions for up to 6 weeks in culture. The reaggregated kidney tubules frequently reached a diameter of 10 mm, but their thickness never exceeded 0.3–0.5 mm. These dimensions were sufficient to permit diffusion of gases and nutrients to all cells in the aggregates; therefore, central necrosis of the tissue did not occur.

These reaggregate culture techniques have been used to study type II alveolar pneumocytes isolated from fetal lung (Douglas *et al.*, 1976; Douglas and Teel, 1976). These investigations have demonstrated that fetal rat lung can be enzymatically dissociated into a monodisperse population of cells and that these cells are capable of reaggregating to form histotypic elements that resemble the epithelial tubules of intact fetal lung. Type II cells in these cultures continue to mature and, after several days *in vitro*, fully differentiated alveolar-like structures (ALS) are formed (Douglas *et al.*, 1976b; Douglas and Teel, 1976). During culture, the glycogen deposits are depleted and lamellar bodies appear in the cytoplasm of these fetal type II cells. The surfactant-containing lamellar bodies are secreted by the type II cells and liberated into the lumen of the ALS. Surfactant material can be isolated from cultures by sucrose density centrifugation (Frosolono *et al.*, 1970). The phospholipid profiles obtained from surfactant fractions of organotypic cultures are similar to the phospholipid profiles obtained from surfactant fractions derived from whole lung (Engle *et al.*, 1979).

This chapter describes the *in vitro* formation, long-term maintenance, and maturation of histotypic structures derived from enzymatically dissociated fetal human lung cells cultured on a gelatin sponge substrate. Biochemical data are presented for cultures established from fetal rat lung when comparable data are lacking from the human cultures.

II.   Materials and Methods

A.   Preparation of Monodisperse Cells

Human fetal lung (18–20 weeks of gestation) was obtained following elective abortion. Under sterile conditions, large airways and pulmonary vessels were removed, leaving tissue consisting primarily of lung parenchyma. This tissue was gently minced into 1-mm³ fragments and transferred to an erlenmeyer flask. The mince was washed three times with RPM1-1640 tissue culture medium, and then an enzyme solution consisting of 0.1% collagenase (type I, Worthington Biochemical Company), 0.1% trypsin (1:250 Difco), and 1% chicken serum (Gibco Company) in calcium- and magnesium-free Hanks' saline was added. The dissociation procedure consisted of a series of 15-minute incubations at 37°C. After each incubation, the cell suspension was harvested and filtered through a 41-μm nylon screen (Nitex HC3-41, Tetko, Inc.) into a 50-ml centrifuge tube on ice. An equal volume of chilled culture medium supplemented with 10% selected fetal bovine serum and antibiotics (antibiotic–antimycotic mixture 100×, Gibco Company, 10 ml/liter) was added, and the cells were stored on ice. Upon completion of the dissociation procedure the cell suspension was centrifuged at 220 g for 6 minutes and the pellet resuspended in serum-supplemented medium. Cell viability was assayed by erythrosine B dye exclusion. Cell viabilities of 70–80% were routinely achieved. The cells were centrifuged a second time (220 g, 6 minutes), and the cell pellet was incubated for 1 hour at 37°C. Fetal rat lung from day 19 of gestation was prepared as described previously (Douglas and Teel, 1976; Engle *et al.*, 1979). Cell viabilities of 80–90% are obtained under similar conditions when fetal rat lung is used.

B.   Culture Procedures

After the incubation, the cells were resuspended in a volume of culture medium to yield 1.0×10^7 viable cells per 50 μl. The suspension was then inoculated (50-μl aliquots) onto the surface of individual 2-cm^2 pieces of media-hydrated Gelfoam collagen sponge (Upjohn). Two cultures were placed in a 100-mm culture dish. The cultures were incubated at 37°C in a humidified atmosphere of 5% carbon dioxide in air for 1 hour to allow the cells to attach to the substrate. Culture medium (20 ml) was added, and the cultures incubated for 48 hours. At this time they were placed on a rocker platform, set at 3 cycles per minute, for the duration of the experiment. Culture medium was replaced every other day.

C.   Glucose Uptake

Media from organotypic cultures (maintained as two squares per dish in 20 ml of RPM1-1640 medium) was sampled at the time of feeding and 20 hours later. The medium was deproteinated by the method of Somogyi (1945), and the glucose concentration assayed with glucose oxidase (Hugget and Nixon, 1975). Uptake was calculated from the difference between initial and final glucose concentrations and expressed as nanomoles of glucose per hour per 10^6 cells. The DNA content of the cultures was measured utilizing mithramycin (Hill and Whatley, 1975).

D.   Phospholipid Analysis

Samples of freshly obtained fetal lung (approximately 1 gm) were homogenized in chloroform–methanol (2:1), washed according to Radin (1969), and separated into phospholipid classes as described by Engle *et al.*, (1976). Surfactant was isolated from organotypic cultures using a modification of the method of Frosolono *et al.* (1970) as described by Sanders and Longmore (1975), with cellular components not associated with surfactant being pooled as a residual fraction. Phospholipids of the surfactant and residual fractions were separated as above. Phospholipids were quantitated by phosphorus assay (Dittmer and Wells, 1969).

E.   Electron Microscopy

Samples were fixed in 2.5% glutaraldehyde in 0.1 M cacodylate buffer (pH 7.3) for 1 hour and postfixed for 30 minutes in 1% osmium tetroxide in 0.1 M cacodylate buffer. Samples were dehydrated in a graded series of acetone in water and embedded in Epon–Araldite. Thin sections were cut with a diamond knife on a Porter–Blum MT-2 ultramicrotome, stained with lead citrate, and examined and photographed in a Philips EM200 electron microscope.

III.   Results

A.   Morphology and Phospholipid Composition of Human Fetal Lung at 18–20 Weeks of Gestation

The human fetal lung at 18–20 weeks of gestation contains epithelial tubules instead of pulmonary alveoli. The tubules are comprised of glycogen-rich epithelial cells containing very few lamellar bodies (Fig. 1). At this stage of lung development microvilli are not present on the apical cell surfaces of these undifferentiated cells (Fig. 1).

The phospholipid composition of human fetal lung at 20 weeks of gestation is shown in Table I. PC (56%) is the predominant phospholipid, with PE (25%) the second most abundant. This is very similar to what is found in 16- to 20-day fetal rat lung (Table I). The low content of PG clearly indicates an immature lung.

B.   Reaggregation of Monodisperse Human Fetal Lung Cells to Form Histotypic Structures

Cells dissociated from human fetal lung are monodisperse when inoculated onto the Gelfoam sponge matrix. During the initial 48 hours of culture, these single cells reaggregate and form structures resembling the epithelial tubules of intact fetal lung. Initially, these tubules are 50–100 μm in diameter, however, by the end of the first week *in vitro,* they increase to 200–400 μm in diameter. The reaggregated tubules are surrounded by fibroblast-like cells which form one to several cell layers immediately subjacent to the epithelium. These spindle-shaped cells also comprise a loose mesenchymal stroma between the epithelial tubules and the trabeculae of the Gelfoam matrix.

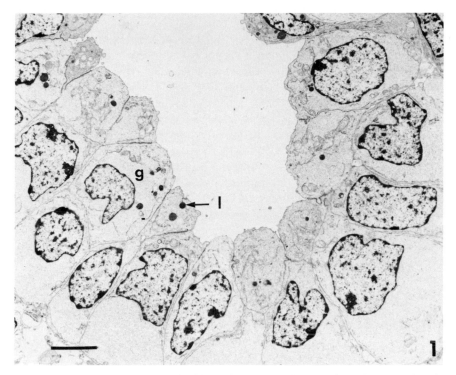

FIG. 1.   Survey electron micrograph of human fetal lung at 20 weeks of gestation. A portion of a single epithelial tubule is illustrated. At this stage of fetal lung development, the epithelial cells are relatively undifferentiated and contain large pools of glycogen (g), few organelles, and a small number of lamellar bodies (1). Apical microvilli are absent. ×3000. Bar = 0.25 μm.

TABLE I

PHOSPHOLIPID COMPOSITION OF 20-WEEK HUMAN
FETAL LUNG AND 16- TO 20-DAY
FETAL RAT LUNG

Phospholipid	Human (%)[a]	Rat (%)[b]
Phosphatidylglycerol	1.01 ± 0.28	1.6 ± 0.2
Phosphatidylethanolamine	25.74 ± 5.53	24.6 ± 0.5
Phosphatidylserine and phosphatidylinositol	1.86 ± 0.43	8.9 ± 1.0
Phosphatidylcholine	56.41 ± 4.19	57.5 ± 1.0
Sphingomyelin	13.25 ± 1.82	5.2 ± 0.8
Others	1.73 ± 0.51	2.2 ± 0.1

[a] Mean ± S.E.M.; n = 6 lungs.
[b] Mean ± S.E.M.; n = 5 litters.

C. Fine Structure of the Reaggregate Culture

Initially, the reaggregated tubules are lined by an undifferentiated epithelium similar to that present in fetal human lung at 18–20 weeks of gestation (see Fig. 1). The epithelial cells have few organelles and contain large amounts of glycogen. As these cells differentiate *in vitro,* they develop characteristics of type II alveolar cells. The epithelial cells assume a cuboidal shape, microvilli develop on the apical cell surface, glycogen stores are depleted, and lamellar bodies appear in the cytoplasm. Figure 2 illustrates a portion of one epithelial tubule lined with numerous type II cells. Microvilli are present on the apical cell surfaces, and they are similar in form and distribution to microvilli present on type II cells of intact lung. Intracellular glycogen deposits are significantly reduced in the cells after 10–14 days in culture (Fig. 2). During this period of *in vitro* cultivation, numerous lamellar bodies appear in the cytoplasm of the type II cells (Figs. 2 and 3). At higher magnification, portions of three type II cells are illustrated in Fig. 3. A well-developed Golgi complex, glycogen deposits, lamel-

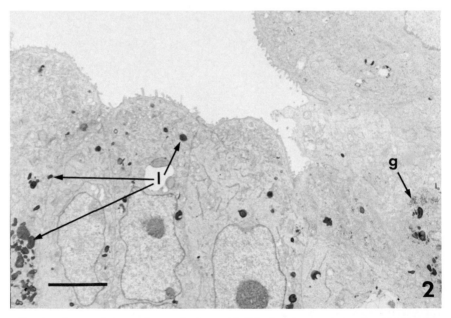

FIG. 2.   Survey electron micrograph of an epithelial tubule present in an organotypic culture established from human fetal lung at 20 weeks of gestation. The culture was maintained 14 days *in vitro* prior to fixation. The tubule is lined with type II cells and, at this stage of *in vitro* development, increased numbers of lamellar bodies (l) are present in the cytoplasm. Glycogen deposits (g) are reduced in the cells, and numerous mitochondria and profiles of RER are found in the apical cytoplasm. Microvilli are present on the apical surfaces of the cells in a distribution pattern similar to that present in whole lung. ×3800. Bar = 0.25 μm.

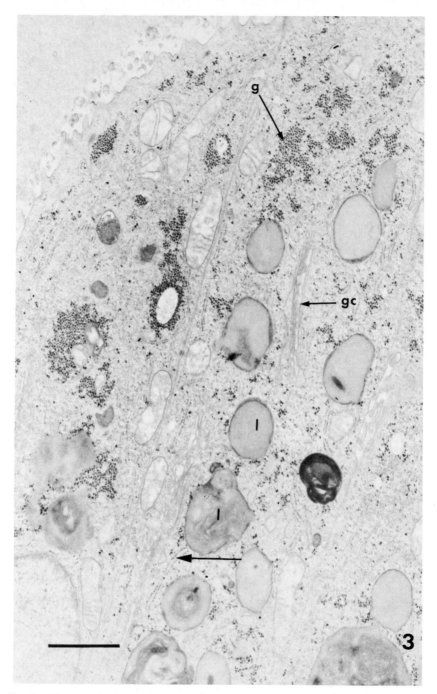

FIG. 3. Electron micrograph of portions of three type II cells in an epithelial tubule present in an organotypic culture after 14 days *in vitro*. Well-developed Golgi complexes (gc) are present in the cells. Small aggregates of glycogen (g) remain in the cytoplasm of the type II cells, and large numbers of lamellar bodies (l) are present. Mitochondria and profiles of RER (arrow) are observed in the cytoplasm, and microvilli are present on the cells' apical surfaces. ×18,400. Bar = 1.0 μm.

lar bodies, and profiles of rough endoplasmic reticulum (RER) are present in the cytoplasm. Surface microvilli are also present. Figure 4 is an electron micrograph illustrating type II cells bordering the lumen of an ALS. Microvilli project from the apical cell surface into the lumen. Lamellar bodies present in these cells have two different appearances—some have a lamellar substructure, while others have a homogeneous appearance. Lamellar bodies displaying an intermediate morphology are also present. A distinct basal lamina is present between these type II cells and the subjacent fibroblast-like cells (Fig. 5).

D. Phospholipid Composition of Organotypic Cultures Established from Fetal Rat Lung

It has also been demonstrated that organotypic cultures established from fetal rat lung continue to differentiate *in vitro* (Douglas *et al.*, 1976b; Douglas and

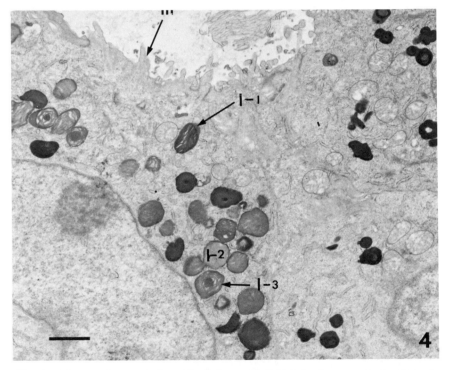

FIG. 4.   Electron micrograph of type II cells in an epithelial tubule of an organotypic culture after 14 days *in vitro*. Microvilli (m) project from the surface of the type II cell into the lumen of the tubule. Numerous lamellar bodies are present in the cytoplasm of each type II cell. The morphology of these organelles varies—some have a lamellar substructure (l-1), and others have a homogeneous appearance (l-2); intermediate forms are also observed (l-3). ×10,400. Bar = 1.0 μm.

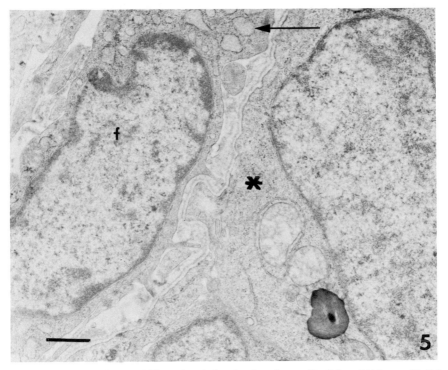

FIG. 5. Electron micrograph illustrating the basal region of a type II cell (asterisk) in an epithelial tubule of an organotypic culture after 14 days *in vitro*. A basal lamina is present between the type II cell and a subjacent fibroblast-like cell (f). A lamellar body, profiles of RER, and mitochondria are present. The RER (arrow) in the fibroblast-like cell is dilated, and the cisternae contain a flocculent material. ×11,400. Bar = 1.0 μm.

Teel, 1976). After 1 week in culture these rat type II cells have proven to be a useful system for studying surfactant synthesis (Engle *et al.*, 1979). With the procedure of Frosolono *et al.* (1970), a surfactant fraction can be isolated from these cultures that is similar to surfactant isolated from adult rat lung. The phospholipid composition of surfactant isolated from organotypic cultures is compared to that isolated from adult rat lung in Table II. We routinely save all cellular material not isolated with the surfactant band as a residual fraction. The residual serves as an internal control when studying surfactant synthesis by type II cells in culture. This is important because most cells in culture synthesize PC, many at rates much greater than that of the type II cell.

PC (71%) is the predominant phospholipid in the surfactant fraction derived from organotypic cultures and is similar to the 69% found in the surfactant fraction derived from adult lung. The content of PE and sphingomyelin is low (9 and 7%, respectively). In contrast, the amount of PC is only 46 and 50% in the

TABLE II

PHOSPHOLIPID COMPOSITION OF ADULT RAT LUNG AND ORGANOTYPIC CULTURES
FROM FETAL RAT LUNG

	Rat lung[a]		Organotypic cultures[b]	
Phospholipid	Surfactant (%)	Residual (%)	Surfactant (%)	Residual (%)
Phosphatidylglycerol	6.7 ± 0.3	2.8 ± 0.2	4.0 ± 0.5	2.2 ± 0.2
Phosphatidylethanolamine	9.8 ± 0.4	23.6 ± 0.7	8.8 ± 1.8	14.2 ± 0.8
Phosphatidylserine and phosphatidylinositol	6.1 ± 0.5	10.5 ± 0.7	5.6 ± 0.5	7.5 ± 0.6
Phosphatidylcholine	68.8 ± 1.3	46.2 ± 1.4	71.5 ± 3.0	50.6 ± 1.7
Sphingomyelin	5.0 ± 0.2	12.5 ± 0.5	6.6 ± 0.5	19.9 ± 1.7
Others	3.6 ± 0.2	4.3 ± 0.5	3.7 ± 0.4	7.6 ± 0.7

[a] Mean ± S.E.M.; $n = 30$.
[b] Mean ± S.E.M.; $n = 44$.

two residual fractions. The residual fraction contains greater amounts of PE (23.6 and 14.2%) and sphingomyelin (12.5 and 19.9%) than the surfactant fraction. Thus, the phospholipid composition of the residual fraction resembles the composition of the membranes of several other organs.

The amount of PG (4%) in the surfactant fraction of organotypic cultures is increased over what is found in the residual fraction (2%), but is not as high as the amount found in the surfactant fraction of adult lung (7%) or lamellar bodies (12%).

Fatty acid analysis of surfactant PC from organotypic cultures demonstrates that myristate (9%), palmitate (79%), and stearate (8%) are the major fatty acids present. The amounts of myristate, palmitate, and stearate are 2.5, 85, and 5%, respectively, in surfactant PC of an adult rat lung (Sanders and Longmore, 1975). Thus, the PC of the surfactant fraction from organotypic cultures is highly saturated, a prime criterion for its function *in vivo*.

Organotypic cultures of rat and human cells are viable for 1–4 weeks as judged by staining with the vital dye tetrazolium chloride (unpublished observation). Glucose uptake by the cultures has been measured as an index of metabolic activity and integrity. Rat cultures use glucose in a concentration-dependent manner with rates of 121, 226, and 300 nmoles/hour/10^6 cells at glucose concentrations of 1.5, 5.6, and 10 mM, respectively (Engle *et al.*, 1979). Insulin stimulates the uptake of glucose and the oxidation of glucose to carbon dioxide by rat cells (Sanders *et al.*, 1979). In contrast, we obtained a glucose uptake of 103 nmoles/hour/10^6 cells at 11 mM glucose after 5 days in culture for human cells. Rat cultures utilize glucose, palmitate, choline, and acetate for the synthesis of surfactant PC and leucine for the synthesis of surfactant proteins (Engle *et al.*, 1979).

A surfactant fraction can also be obtained from organotypic cultures of human lung cells after 7–10 days of culture. Again, the material is enriched in PC. However, insufficient samples have been obtained at this time to report an analysis.

IV. Discussion and Perspectives

The organotypic culture system was developed for the study of maturation of type II cells and for the study of surfactant synthesis. This system has several advantages over other lung cell culture methods presently in use. The cells in organotypic culture, unlike those in mixed lung cell cultures or explants, are predominantly type II cells (85–95%), thereby greatly reducing the complication of non-surfactant-producing cells. In contrast to cells in monolayer systems, type II cells in organotypic cultures retain the alveolar structure of the intact lung, thus retaining the structural integrity, cell-to-cell contact, and polarity of type II cells *in vivo*. This polarity includes the presence of a mesenchymal layer on the basal side. Smith (1979) has proposed that lung fibroblasts secrete a factor that enhances surfactant synthesis by type II cells. The alveolar-like architecture and mesenchymal basal layer may be essential, since this is the only system reported to date that permits the maintenance of viable type II cells for more than 4–5 days in culture.

Organotypic cultures may be established from fetal lung at various periods of gestation. If cultures are established from early fetal lung, the undifferentiated cells present in the reaggregated epithelial tubules will mature *in vitro*. These developing type II cells lose their glycogen deposits and develop microvilli on their apical cell surface, and lamellar bodies appear in the cytoplasm. These surfactant-containing organelles are secreted by the type II cells into the lumen. Examination of human fetal lung organotypic cultures at this critical period in fetal lung maturation may yield information concerning the induction of surfactant synthesis, data that are unobtainable using freshly isolated adult type II cells. In addition, since these cultures can be maintained for up to 4 weeks *in vitro*, the full effects of mediators on surfactant synthesis may be studied. Finally, since surfactant and residual fractions can be isolated from these type II cells, a direct analysis of events involved in the metabolism of surfactant phospholipids and proteins is possible.

The surfactant and residual fractions from organotypic cultures of type II cells derived from fetal rat lung have phospholipid profiles similar to those of the surfactant and residual fractions from adult rat lung (Engle *et al.*, 1979). However, there are some minor differences. PG, a proposed marker for surfactant, is present in smaller percentages in the surfactant fraction derived from cultures (4.0%), as compared to surfactant fractions derived from whole lung (6.7%). This difference may be related to lung maturity, since PG does not reach substantial concentrations in the rat lung until near the time of birth. The residual fraction from the isolated cells has slightly more PG (4.2 versus 2.8%) and slightly less PE (14.4 versus 23.6%) than the whole-lung residual fraction. These differences may be a result of the cellular heterogeneity of adult lung tissue.

Glucose uptake by rat and human lung cells in organotypic cultures was measured as an index of metabolic integrity. Glucose uptake by the rat cells, as measured by glucose disappearance from the medium, was 226 nmoles/hour/10^6 cells when the medium glucose concentration was 5.6 mM. The glucose uptake by these cells was 10 times that reported by Mason *et al.* (1977b) for freshly isolated type II cells. Mason measured the glucose uptake by type II cells within a few hours after the cells were isolated by trypsinization, in contrast to our measurements after 8 days in culture. Proteolytic enzymes have been reported to alter cellular metabolism and inhibit phospholipid synthesis in alveolar macrophages (Lindenmayer *et al.*, 1968). Thus, the type II cells used by Mason *et al.* (1977b) may not have fully recovered from the deleterious effects of trypsinization. Mason *et al.* (1977b) found that alveolar macrophages that had not been isolated by trypsin treatment took up glucose at a rate of 127 nmoles/hour/10^6 cells when the medium glucose was 2.7 mM. This is similar to the rate of 121 nmoles/hour/10^6 cells we found with organotypic cultures of type II cells incubated in 1.5 mM glucose (Engle *et al.*, 1979). The difference in glucose uptake by type II cells in the two systems may also be an age-related metabolic difference, since Mason *et al.* (1977b) employed cells from adult animals while we utilized a maturing cell system.

We have not yet measured the glucose uptake by organotypic cultures of human type II cells after 12–15 days *in vitro* when they are morphologically mature. However, they consume glucose at a rate of 103 nmoles/10^6 cells/hour after 5 days *in vitro*. This is lower than the rate found for rat cultures and may reflect a lower metabolic rate in human cells compared to rat cells. Alternatively, the lower rate could be due to the fact that the measurement was made at a time when the human cells were still depleting their glycogen stores.

The organotypic system has proven to be useful for studying both differentiation and surfactant synthesis in type II cells from fetal rat lung. The present chapter describes maturation of human type II cells. Studies on surfactant synthesis in human cells remain to be performed. It is important to point out that the process of placing human fetal lung in culture accelerates the maturation process. Tissue that still has to undergo 20 weeks of *in vivo* development matures in 2 weeks of *in vitro* culture. It remains to be seen whether hormones that stimulate maturation *in vivo,* i.e., glucocorticoids and thyroid hormones, can further accelerate maturation in these cultures of human type II cells.

ACKNOWLEDGMENT

We thank Mrs. Janice Bonaccorso for excellent technical assistance.

REFERENCES

Adamson, I. Y. R., and Bowden, D. H. (1973). *Exp. Mol. Pathol.* **18,** 112–124.
Ansevin, K. D., and Lipps, B. U. (1973). *In Vitro* **8,** 483–488.
Askin, F. B., and Kuhn, C. (1971). *Lab. Invest.* **25,** 260–268.
Batenburg, J. J., Longmore, W. J., and Van Golde, L. M. G. (1978). *Biochim. Biophys. Acta* **529,** 160–170.
Dermer, G. B. (1970). *J. Ultrastruct. Res.* **33,** 306–317.
Dittmer, J. C., and Wells, M. A. (1969). *In* "Methods in Enzymology" (J. M. Lowenstin, ed.), Vol. 14, pp. 482–530. Academic Press, New York.
Douglas, W. H. J., and Teel, R. W. (1976). *Am. Rev. Respir. Dis.* **113,** 17–23.
Douglas, W. H. J., Farrell, P. M., and O'Hare, K. H. (1976a). *Am. Rev. Respir. Dis.* **113,** 209.
Douglas, W. H. J., Moorman, G. W., and Teel, R. W. (1976b). *In Vitro* **12,** 373–381.
Engle, M. J., Sanders, R. L., and Longmore, W. J. (1976). *Arch. Biochem. Biophys.* **173,** 586–595.
Engle, M. J., Sanders, R. L., and Douglas, W. H. J. (1980). *Biochim. Biophys. Acta* (in press).
Felts, J. M. (1964). *Health Phys.* **10,** 973–979.
Fischman, D. A., and Moscona, A. A. (1971). *In* "Cardiac Hypertrophy" (N. R. Alpert, ed.), pp. 125–139. Academic Press, New York.
Frosolono, M. F., Charms, B. L., Pawlowski, P., and Slivka, S. (1970). *J. Lipid Res.* **11,** 439–457.
Frosolono, M. F., Kress, Y., Wintter, M., and Rosenbaum, R. M. (1976). *In Vitro* **12,** 708–717.
Funkhouser, J. D., and Hughes, E. R. (1977). *J. Steroid Biochem.* **8,** 519–524.
Garber, B. B., and Moscona, A. A. (1972). *Dev. Biol.* **27,** 217–234.
Gil, J., and Reiss, O. K. (1973). *J. Cell Biol.* **58,** 152–171.
Godinez, R. I., Sanders, R. L., and Longmore, W. J. (1975). *Biochemistry* **14,** 830–834.
Gross, I., Walker-Smith, G. J., Maniscalco, W. M., Czajka, M. R., Wilson, C. M., and Rooney, S. A. (1978). *J. Appl. Physiol.* **45,** 355–362.
Hallman, M., Miyai, K., and Wagner, R. M. (1976). *Lab. Invest.* **35,** 79–86.
Hill, B. T., and Whatley, S. (1975). *FEBS Lett.* **56,** 20–23.
Hugget, A., and Nixon, D. (1975). *Lancet* **2,** 368–370.
Kalifat, S. R., Dupuy-Coin, A. M., and Delarue, J. (1970). *J. Ultrastruct. Res.* **32,** 572–589.
King, R. J. (1974). *Fed. Proc., Fed. Am. Soc. Exp. Biol.* **33,** 2238–2247.
King, R. J. (1977). *Am. Rev. Respir. Dis.* **115,** 73–81.
King, R. J., and Clements, J. A. (1972). *Am. J. Physiol.* **223,** 715–726.
Lindenmayer, G. E., Sordahl, L. A., and Schwartz, A. (1968). *Circ. Res.* **23,** 439–450.
Mason, R. J., Williams, M. C., and Dobbs, L. G. (1977a). *ERDA Symp. Ser.* **43,** Conf-760927, p. 280.
Mason, R. J., Williams, M. C., Greenleaf, R. D., and Clements, J. A. (1977b). *Am. Rev. Respir. Dis.* **115,** 1015–1026.

Massaro, G. D., and Massaro, D. (1972). *Am. Rev. Respir. Dis.* **105**, 927–931.
Meyrick, B., and Reid, L. M. (1977). *In* "Development of the Lung" (W. A. Hodson, ed.), Vol. 6, pp. 135–214. Dekker, New York.
Mims, L. C., Mazzuckelli, L. F., and Kotas, R. V. (1975). *Pediatr. Res.* **9**, 165–167.
Moscona, A. A. (1962). *J. Cell. Comp. Physiol.* **60**, 65–80.
Moscona, A. A., and Moscona, H. (1952). *J. Anat.* **86**, 287–301.
Paul, G. W., Hassett, R. J., and Reiss, O. K. (1977). *Proc. Natl. Acad. Sci. U.S.A.* **74**, 3617–3620.
Radin, N. S. (1969). *In* "Methods in Enzymology" (J. M. Lowenstein, ed.), Vol. 14, pp. 245–254. Academic Press, New York.
Sanders, R. L., and Longmore, W. H. (1975). *Biochemistry* **14**, 835–840.
Sanders, R. L., Engle, M. J., and Douglas, W. H. J. (1979). *Fed. Proc. Fed. Am. Soc. Exp. Biol.* **38**, 1436.
Sankaran, L., Proffitt, R. T., Peterson, J. R., and Pogell, B. M. (1977). *Proc. Natl. Acad. Sci. U.S.A.* **74**, 4486–4490.
Smith, B. T. (1979). *Science* **204**, 1094–1095.
Smith, B. T., and Torday, J. S. (1974). *Pediatr. Res.* **8**, 845–851.
Smith, F. B., and Kikkawa, Y. (1978). *Lab. Invest.* **38**, 45–51.
Somogyi, M. (1945). *J. Biol. Chem.* **160**, 69–73.
Sorokin, S. (1961). *Dev. Biol.* **3**, 60–83.
Sorokin, S. P. (1970). *AEC Symp. Ser.* **21**, 3–41.
Spitzer, H. L., Nekola, M. V., Porter, J. C., and Johnston, J. M. (1975). *Gynecol. Invest.* **6**, 39–40.
Van Golde, L. M. G. (1978). *In* "Lung Diseases: State of the Art" (J. F. Murray, ed.), pp. 375–398. Am. Lung Assoc., New York.
Williams, M. C. (1977). *J. Cell Biol.* **72**, 260–277.
Wykle, R. L., Malone, B., and Snyder, F. (1977). *Arch. Biochem. Biophys.* **181**, 249–256.

Chapter 6

The Human Alveolar Macrophage

GARY W. HUNNINGHAKE, JAMES E. GADEK,
SUSAN V. SZAPIEL, IRA J. STRUMPF, OICHI KAWANAMI,
VICTOR J. FERRANS, BRENDAN A. KEOGH,
AND RONALD G. CRYSTAL

Pulmonary Branch and Pathology Branch,
National Heart, Lung, and Blood Institute,
National Institutes of Health,
Bethesda, Maryland

I. Introduction

Alveolar macrophages are part of a family of mononuclear phagocytic cells that are widely scattered throughout the body (Langevoort *et al.*, 1970). Besides alveolar macrophages, this family includes blood monocytes, Kupffer cells of the liver, lining macrophages of the spleen and bone marrow sinusoids, free mac-

ISBN 0-12-564121-4

(Model FN, Coulter Electronics, Hialeah, Florida), viability (trypan blue dye exclusion), and differential count (Wright–Giemsa-stained cytocentrifuge preparation; Shanden Southern Instruments, Sewickley, Pennsylvania) and resuspended in HBSS at the desired cell density.

Alveolar macrophages and lung lymphocytes can be obtained directly from lung biopsy specimens by placing the specimens immediately in sterile heparinized saline at 4°C at the time of thorocotomy. The tissue is then transferred to a petri dish containing HBSS, and the parenchyma is teased using a forceps and a surgical scalpel blade. The resulting material is filtered through several layers of sterile surgical gauze to obtain a single-cell suspension. Mononuclear cells (alveolar macrophages and lymphocytes) are then isolated from these cell suspensions by Hypaque–Ficoll density centrifugation. The Hypaque–Ficoll centrifugation procedure separates alveolar macrophages and lymphocytes from other lung parenchymal cells present in the unfractionated biopsy cell suspension.

B.   Methods for Culturing Alveolar Macrophages

Human alveolar macrophages (obtained by bronchoalveolar lavage or from lung biopsy specimens) can be maintained in culture for at least 60 days in an appropriate environment. Although alveolar macrophages can be successfully cultured by a variety of different methods, these cells exhibit various unique functional and morphological characteristics which must be considered in choosing the optimal culture method for the type of *in vitro* studies to be carried out.

One of the critical determinants of the ability of alveolar macrophages to function and/or survive is whether they are attached to a surface or are suspended in a fluid. Macrophages attached to a surface synthesize greater amounts of protein for longer periods of time and are more efficient at phagocytosis than suspended cells (Leffingwell and Low, 1975) (Table III). However, while such studies demonstrate that the function of alveolar macrophages can be manipulated by providing a surface on which these cells can function, it is not clear that *in vitro* studies utilizing adherent macrophages are more physiological than those utilizing cells in suspension. It is likely that neither system approximates the normal environment of these cells; therefore, considerable care must be taken in extrapolating results from these *in vitro* situations. It is clear, however, that the survival of these cells in culture is very poor unless they can attach to a surface. Therefore, they must be cultured in chambers that have a flat surface to which they can attach.

Alveolar macrophages, in contrast to other macrophages such as Kupffer cells and peritoneal macrophages, normally function in an environment where the partial pressure of oxygen approximates 100 mm Hg. Because they have adapted themselves to this aerobic environment, these cells utilize significantly greater

TABLE III

RELATIVE FUNCTIONAL CHARACTERISTICS OF SUSPENDED AND ADHERENT
MACROPHAGES[a]

Alveolar macrophages	Protein biosynthesis[b]		Phagocytosis[c]	
	60 minutes	180 minutes	10 minutes	30 minutes
Adherent	2+	4+	2+	4+
Nonadherent	2+	2+	2+	2+

[a] Adapted from Leffingwell and Low (1975).
[b] Time course of incorporation of radioactive leucine into protein by alveolar macrophages.
[c] Time course of oil red O-albumin particle uptake by alveolar macrophages.

quantities of oxygen and demonstrate a lower rate of glycolysis than peritoneal macrophages (Simon et al., 1977). The increased level of oxygen consumption by alveolar macrophages compared to that of peritoneal macrophages appears to be due to the increased levels of enzymes involved in oxidative phosphorylation and to the decreased amounts of glycolytic enzymes used in glycolysis (Table IV). The difference in bioenergetics between alveolar macrophages and peritoneal macrophages is short-lived in culture, however, since both of these cell populations can readily alter the relative amounts of their enzymes involved in oxidative phosphorylation and glycolysis to adapt to a wide range of oxygen tensions in their in vitro environment (usually within 96 hours). These studies demonstrate that the partial pressure of oxygen may have profound effects on the energy-dependent functions (such as phagocytosis and pinocytosis) of the alveo-

TABLE IV

RELATIVE QUANTITIES OF ENZYMES INVOLVED WITH
OXIDATIVE PHOSPHORYLATION AND GLYCOLYSIS IN
ALVEOLAR MACROPHAGES COMPARED TO PERITONEAL
MACROPHAGES[a]

Type of macrophage	Oxidative phosphorylation[b]	Glycolysis[c]
Alveolar	4+	1+
Peritoneal	2+	4+

[a] Adapted from Simon et al. (1977).
[b] Activity of cytochrome oxidase.
[c] Activity of pyruvate kinase or phosphofructokinase.

lar macrophage, and this must be carefully considered in planning *in vitro* studies. In addition, when the partial pressure of oxygen is abruptly changed in the environment of these cells, a period of adaptation (up to 96 hours) may be required before the macrophages function "normally."

Alveolar macrophages are intrinsically very metabolically active cells (Simon *et al.*, 1977). For example, they utilize 1.5 μmoles of oxygen per 10^6 cells per hour, which is up to four times as much oxygen per cell as cultured lung fibroblasts utilize (Bradley *et al.*, this volume). As a result of this activity, these cells rapidly deplete essential nutrients from and add large amounts of lactic acid to the culture medium. In addition, the functional capacity of alveolar macrophages appears to increase with time in culture. As shown by Cohen (1973), after 3–4 weeks of culture, human alveolar macrophages "differentiate" into multinucleated giant cells. Such cells may have up to 30 times the phagocytic capacity (per cell) and are more active metabolically than freshly isolated alveolar macrophages. For these reasons the culture medium of alveolar macrophages must be supplemented or replaced periodically; the frequency with which this is done depends on the length of time the cells have been in culture.

The following methods are used to culture alveolar macrophages in our laboratory (Fig. 1). For long-term cultures (>24 hours), the cells are placed in petri dishes or multichambered, flat-bottom plates in RPMI-1640 medium with 20%

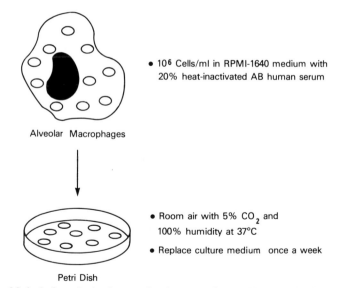

Alveolar Macrophages

● 10^6 Cells/ml in RPMI-1640 medium with 20% heat-inactivated AB human serum

● Room air with 5% CO_2 and 100% humidity at 37°C

● Replace culture medium once a week

Petri Dish

Fig. 1.   Methods for culturing human alveolar macrophages. The macrophages are cultured in petri dishes at a density of 10^6 cells/ml in RPMI-1640 with 20% heat-inactivated AB human serum in room air with 5% carbon dioxide and 100% humidity at 37°C. The culture medium is replaced once a week.

heat-inactivated human AB serum at a density of 10^6 cells/ml in room air with 5% carbon dixoide and 100% humidity at 37°C. The medium is usually replaced once a week. If the cells are maintained in culture for periods of time greater than 21 days, then the culture medium is replaced twice per week.

Without serum, human alveolar macrophages remain alive and functionally active in culture for 24–48 hours. After this time the viability of these cells begins to drop sharply. If the planned *in vitro* study cannot be performed within the first 24–48 hours of culture or in the presence of serum, the cells are initially placed in culture with serum. After a period of time, the serum is removed and replaced with serum-free medium; the cells will then remain viable in RPMI-1640 for up to 48 hours. With the use these methods, various functional aspects of alveolar macrophages can be studied in a relatively serum-free system at any point in time following initiation of the cultures.

III. Results

A. Characteristics of Cell Suspensions Obtained from Bronchoalveolar Lavage Fluid and Lung Biopsy Specimens

The cell yield of bronchoalveolar lavage in normal nonsmoking individuals varies from 5×10^6 to 10×10^6 cells per 100 ml lavage (Table V). Significantly larger numbers of cells are obtained from cigarette smokers and patients with a variety of lung diseases. In our studies of normal nonsmoking volunteers, 93 ± 5% of the cells in lavage fluid were alveolar macrophages (Fig. 2a), and 7 ± 1%

TABLE V

CHARACTERISTICS OF THE MONONUCLEAR CELL SUSPENSION OBTAINED FROM BRONCHOALVEOLAR LAVAGE FLUID AND LUNG BIOPSY SPECIMENS

Method	Yield of macrophages	Parenchymal lung cells (% of cells)[a]	Peripheral blood contamination (% of cells)
Bronchoalveolar lavage	$5–10 \times 10^{6b}$	1	1
Lung biopsy	$4–9 \times 10^{6c}$	1^d	3

[a] Peripheral blood contamination was estimated by quantitating the total number of red blood cells in lavage fluid or in biopsy cell suspensions prior to Hypaque–Ficoll centrifugation.

[b] Cell yield per 100 ml lavage.

[c] Cell yield per cubic centimenter of lung tissue.

[d] Percentage of parenchymal lung cells in biopsy cell suspensions after Hypaque–Ficoll centrifugation.

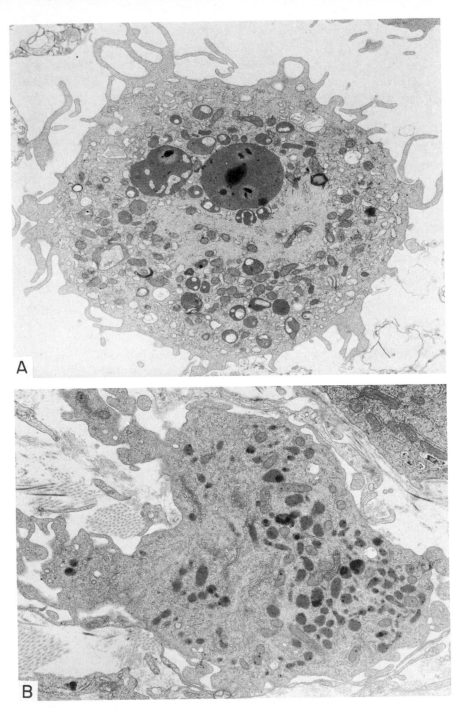

Fig. 2. Transmission electron micrographs of human alveolar macrophages. (A) Normal alveolar macrophage on the epithelial surface of an alveolus. ×9500. (B) Alveolar macrophage in the interstitium of a lung from a patient with idiopathic pulmonary fibrosis. ×12,500.

of the cells were lymphocytes. Polymorphonuclear leukocytes (PMNs) are rarely found in the lungs of normal individuals. Based on the total number of red blood cells in the lavage fluid, the contamination of the lavage fluid cells by cells from peripheral blood is less than 1%.

The cell yield per biopsy varies depending on the size and cellularity of the specimen (Table V). In biopsy specimens from normal lung, the cell yield varies from 4×10^6 to 9×10^6 alveolar macrophages and lymphocytes per cubic centimeter of lung tissue. Larger numbers of cells are obtained from specimens from patients with parenchymal lung disease. The post-Hypaque–Ficoll cell suspensions isolated from biopsies of normal lung contain $85 \pm 5\%$ macrophages (Fig. 2b) and $15 \pm 4\%$ lymphocytes. Based on the total number of red blood cells in the unfractionated biopsy cell suspensions, the contamination of the biopsy cell suspensions by cells from peripheral blood is less than 3%.

B. Removal of Noncellular Substances from Alveolar Surfaces

There is ample evidence that normal human alveolar macrophages have the capacity to ingest a variety of particles deposited in the lower respiratory tract. For example, light microscopic evaluation of alveolar macrophages recovered from cigarette smokers reveals numerous pigmented cytoplasmic granules throughout these cells (Brody and Craighead, 1975). By electron microscopy these pigmented granules appear to represent, at least in part, increased numbers of lysosomes and phagolysosomes. Within these cytoplasmic organelles are disk- or platelike structures, some of which have been identified as kaolinite (Brody and Craighead, 1975). As a result of ingestion of smoke particles, alveolar macrophages of smokers appear to be "activated" in comparison with normal alveolar macrophages (Harris et al., 1970). In comparison with cells obtained from nonsmokers, these macrophages have a higher resting energy requirement, are larger, and contain more Golgi vesicles, endoplasmic reticulum, and residual bodies. In addition, they adhere more rapidly to glass surfaces than normal alveolar macrophages.

Many animal studies have demonstrated that alveolar macrophages have the capacity to ingest a variety of inorganic and organic particles (Green, 1970). Some of these, such as silica particles, are cytotoxic for these cells.

In addition to the clearance of foreign particles from the lower respiratory tract, alveolar macrophages may also participate in the clearance of endogenous substances, such as surfactant (Nichols, 1976), α-antitrypsin (Cohen, 1973), and neutrophil elastase (Campbell et al., 1979), from the alveolar surfaces. As a correlate of these studies, Golde et al. (1976) has demonstrated large amounts of a lipoproteinaceous material, presumably surfactant, in alveolar macrophages of patients with alveolar proteinosis. Alveolar macrophages from these patients demonstrate impaired adherence to glass, depressed response to a chemotactic stimulus, and a defective ability to kill fungi. Such studies suggest that an abnormal environment *in vivo* can markedly affect alveolar macrophage function. In addition, the interaction of alveolar macrophages with antiproteases and proteases may be important in protecting alveolar structures from proteolytic attack.

C.   Phagocytosis and Killing of Microorganisms

One major function of the alveolar macrophage is to maintain the sterility of the lower respiratory tract (Green, 1970). In order for these cells to ingest and kill microorganisms, they must be capable of migrating toward (chemotaxis) and attaching to (adherence) the organism. In this regard, recent studies have demonstrated that normal human alveolar macrophages are capable of responding to various chemotactic stimuli (Warr and Martin, 1974). It is likely, therefore, that these macrophages are capable of directed movement within the alveolar structures. Immunological factors, termed opsonins, are of great importance in promoting the attachment of an alveolar macrophage to particles and microorganisms. These opsonins are primarily immunoglobulins and complement-derived factors. It is known that human alveolar macrophages have receptors for IgG and the C3b fragment of complement (Reynolds *et al.*, 1975) and that both of these opsonins are found on the epithelial surface of the normal human lung (Reynolds and Newball, 1974; Reynolds *et al.*, 1977).

The ability of the alveolar macrophage to ingest and kill microorganisms has been demonstrated *in vivo* and *in vitro* in many types of animal models (Green, 1970; Green and Kass, 1964; Hunninghake and Fauci, 1976b). Studies of human alveolar macrophages *in vitro* have paralleled these animal studies (Cohen and Kline, 1971; Harris *et al.*, 1970). Indeed, the fact that the normal lower respiratory tract is free of viable bacteria is a tribute to the efficiency of the antimicrobial capabilities of these cells, since bacterial contamination by inspired air and aspiration of upper respiratory tract secretion is a common occurrence.

It is clear, therefore, that the alveolar macrophage possesses all the functional capabilities (chemotaxis, adherence, phagocytosis, and killing) necessary to kill most bacteria that reach the alveolar surfaces. To further illustrate these capabilities of the alveolar macrophage, the sequence of attachment, phagocytosis, and killing of antibody-coated red cells is shown in Figs. 3 and 4. These studies demonstrate that normal alveolar macrophages have a greater potential for destroying certain cells and microorganisms than blood monocytes; however, the cytotoxic effector cell function of the alveolar macrophage and the neutrophil are comparable.

Macrophages, including alveolar macrophages, are a major line of host defense against certain viruses. Liu (1955), using a ferret model of influenza infections, demonstrated that alveolar macrophages were capable of inhibiting the growth of certain viruses but not others. Such observations have obvious

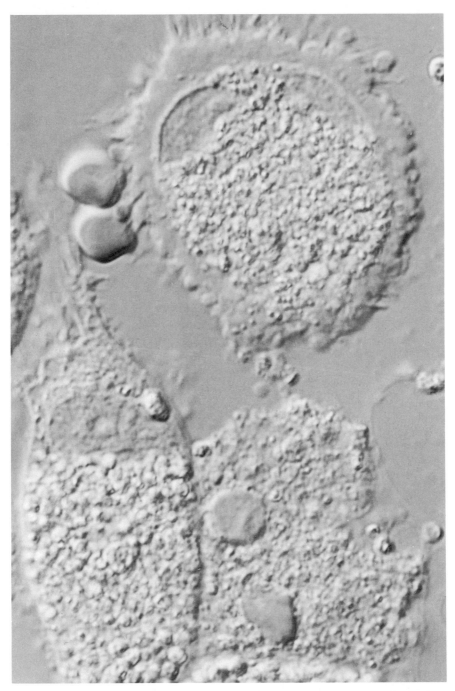

Fig. 3. Nomarski differential interference contrast optics of human alveolar macrophages, demonstrating adherence to and phagocytosis of IgC-coated ox red blood cells. The macrophage on the left has not adhered to or ingested red blood cells. The macrophage at the top has adhered to several red blood cells but has not ingested them. The macrophage on the right has ingested two red blood cells.

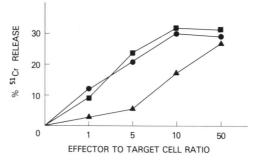

FIG. 4.   A comparison of the ability of human alveolar macrophages (circles), blood neutrophils (squares), and blood monocytes (triangles) to kill ^{51}Cr-labeled, antibody-coated ox red blood cells. The percentage of ^{51}Cr release (percentage of killing of ox red blood cell targets) is on the ordinate, and the ratio of the effector cells to the target cells is indicated on the abscissa.

implications for the pathogenesis of viral infections in humans. In this regard, recent studies by Drew *et al.* (1979) have shown that human alveolar macrophages permit the growth of some strains of herpesvirus but not of cytomegalovirus. One method by which alveolar macrophages may inhibit the growth of certain viruses is via the secretion of interferon (Acton and Myrvik, 1966). In addition, since macrophages possess receptors for the Fc portion of IgG, they are capable of ingesting viruses complexed to antibody.

Some microorganisms, such as *Mycobacterium tuberculosis,* are not killed by alveolar macrophages unless the macrophages have been "activated." Alveolar macrophages can be activated both nonspecifically and specifically to kill these types of organisms (Myrvik and Acton, 1977) but, in general, alveolar macrophages activated via a specific immune reaction have a greater potential for killing than macrophages activated by nonspecific mechanisms. As an example of nonspecific activation of alveolar macrophages, BCG-vaccinated mice are highly resistant to weakly pathogenic organisms such as *Lysteria monocytogenes,* but exhibit no detectable resistance to highly virulent organisms such as *Francisella tularensis.* In contrast, specifically immunized mice are resistant to challenge with *F. tularensis.*

The alveolar macrophage may also function to remove various abnormal cells in its environment. Zwilling and Campolito (1977) have demonstrated that BCG-activated hamster alveolar macrophages are capable of killing virus-transformed tumor cells, suggesting that alveolar macrophages may play a role in killing tumor cells or cells infected with certain viruses.

D.   Secretion of Enzymes

Studies utilizing alveolar macrophages obtained from experimental animals have shown that, following phagocytosis, these cells secrete a variety of enzymes, including neutral proteases, collagenase, and elastase (Condon and Harris, 1978; Horwitz *et al.*, 1976a; Harris *et al.*, 1975; Levine *et al.*, 1976). In contrast, other enzymes, such as lysozyme, are secreted independently of phagocytosis. The amount of each of these enzymes released from alveolar macrophages may be related to the relative state of activation of these cells, since alveolar macrophages of rabbits immunized with BCG *in vivo* release more collagenase *in vitro* than normal cells (Horwitz and Crystal, 1976).

Few studies on enzyme secretion by human alveolar macrophages have been reported. However, studies have demonstrated that alveolar macrophages of smokers, but not normal alveolar macrophages, release elastase *in vitro* (Rodriguez *et al.*, 1977). Although it is clear that normal human alveolar macrophages are capable of secreting a wide variety of enzymes, the release of enzymes from these cells following stimulation *in vitro* differs considerably, in kinetics and in amount, from the release of these same enzymes from neutrophils (Gadek *et al.*, 1979). Neutrophils store significant quantities of collagenase, elastase, and neutral protease, and they rapidly release large quantities of these enzymes following stimulation. In comparison, these enzymes are not secreted by stimulated alveolar macrophages until they have been in culture for 24–48 hours.

In conjunction with their ability to secret various enzymes, alveolar macrophages may also possess inhibitors of these enzymes. Blondin *et al.* (1972) demonstrated an inhibitor of elastase in human alveolar macrophages. In this regard, Cohen (1973) reported by the presence of α1-antitrypsin (a potent inhibitor of elastase) by immunofluorescence in normal human alveolar macrophages and a diminished amount in alveolar macrophages of a subject deficient in serum α1-antitrypsin. In addition, Olsen *et al.* (1975) demonstrated that alveolar macrophages of smokers contained greater amounts of α1-antitrypsin than those of nonsmokers. It is known that much of the α1-antitrypsin in alveolar macrophages is complexed with enzyme (Harris and Goldblatt, 1976), but it is not clear whether the α1-antitrypsin within macrophages is phagocytized as free antiprotease which complexes with protease within the cell or whether such intracellular α1-antitrypsin represents phagocytized preformed protease-antiprotease complexes.

E. Secretion of Mediators

The secretion of a variety of soluble mediators is a well-described function of mononuclear phagocytes. Although little is known about the secretion of these mediators by alveolar macrophages, it is likely that these cells possess similar capabilities. Acton and Myrvik (1966) have shown that rabbit alveolar macrophages inoculated with parainfluenza-3-virus *in vitro* produce a viral inhibitor with properties of interferon. The secretion of interferon by alveolar macrophages may be one means by which these cells inhibit replication of viruses that infect the lower respiratory trace. Rabbit alveolar macrophages are also capable of secreting a lymphocyte-activating factor following stimulation *in vitro* (Ulrich, 1977). The production of this factor by alveolar macrophages may be an important mechanism by which these cells interact with lung lymphocytes in various cell-mediated immune reactions. Golde *et al.* (1974a) showed that human alveolar macrophages were capable of producing a mediator possessing colony-stimulating activity. This factor stimulates the clonal growth of bone marrow cells *in vitro* and may be an important physiological regulator of granulocytopoiesis and monocytopesis *in vivo*. Studies have also demonstrated that alveolar macrophages are capable of secreting a potent low-molecular-weight (400–600) chemotactic factor for neutrophils following *in vivo* or *in vitro* stimulation (Kazmierowski *et al.*, 1977; Hunninghake *et al.*, 1978). The stimuli capable of triggering the release of this chemotactic factor from alveolar macrophages include various microorganisms, organic and inorganic particulates, and immune complexes. Intratracheal injection of the partially purified factor into the lungs of animals results in a rapid accumulation of PMNs within the lung (Hunninghake *et al.*, 1978). The secretion of this factor by alveolar macrophages may be an important means by which these cells can amplify both acute and chronic inflammatory responses characterized by an accumulation of neutrophils within the lung.

F.   Accessory Cells in Immune Responses

The presence of macrophages is necessary for both humoral and cell-mediated immune responses of human lymphocytes to various antigens and mitogens.

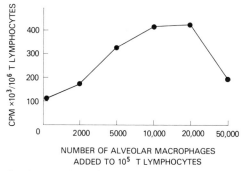

FIG. 5.   The effect of various numbers of alveolar macrophages on the blastogenic response of lymphocytes to concanavalin A (10 μg/ml). Blood T lymphocytes (10*b35* cells) were cultured with concanavalin A on microtiter plates with various numbers of alveolar macrophages in RPMI-1640 medium with 10% AB human serum. Following 5 days in culture the [^{3}H]thymidine incorporation of the labeled lymphocytes was evaluated. The data are expressed as counts per minute times 10^3 of tritiated thymidine incorporated per 10^6 T lymphocytes.

Alveolar macrophages can also function as accessory cells in both antigen- and mitogen-stimulated proliferation of lymphocytes *in vitro* (Lussier *et al.*, 1978; Laughter *et al.*, 1977; Schuyler *et al.*, 1978; Warr and Martin, 1973). Although these studies suggest that one role of the alveolar macrophage is to "present" antigen to lung lymphocytes, it is not clear how such an accessory role relates to the known protective role of alveolar macrophages in excluding antigen from the lung. Independent of the importance of the role of the alveolar macrophage as an accessory cell, *in vitro* studies have demonstrated that optimum mitogen-stimulated lymphocyte proliferation occurs when the ratio of alveolar macrophages to lymphocytes is approximately 1:4 (Fig. 5). When alveolar macrophages comprise greater than 20% of these cultures, lymphocyte proliferation is inhibited. Thus, optimal macrophage–lymphocyte interaction is likely regulated, in part, by the physical relationships of these cells within the alveolar structures.

G.   Origin and Kinetics of the Human Alveolar Macrophage

The ultimate origin of alveolar macrophages is bone marrow (Langevoort *et al.*, 1970). Evidence for the bone marrow origin of human alveolar macrophages was obtained by Thomas *et al.* (1976) in patients with allogenic bone marrow transplants from siblings of the opposite sex. In these studies, alveolar macrophages of the recipient of the transplant were identified as male or female in origin by the presence or absence of a fluorescent Y chromosome. In these

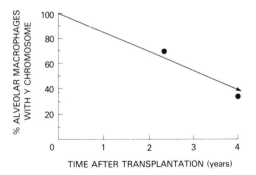

FIG. 6.   The kinetics of replacement of host alveolar macrophages by donor alveolar macrophages following bone marrow transplantation. A male patient with aplastic anemia received a successful bone marrow transplant from his major histocompatibility locus-matched sister. Alveolar macrophages were obtained by bronchoalveolar lavage, and the percentage of host alveolar macrophages was determined utilizing a fluorescent staining technique that detected the presence of a Y chromosome within the macrophages. The percentages of alveolar macrophages with a Y chromosome (host origin) are on the ordinate, and the length of time following bone marrow transplantation is indicated on the abscissa.

patients, the majority of the host macrophages were replaced by macrophages of donor origin in less than 100 days. It is not clear, however, whether all alveolar macrophages on the epithelial surface of the lower respiratory tract were cells that migrated directly to the lung from the blood or whether some of these cells were derived from lung macrophages that replicated *in situ*. Human alveolar macrophages are capable of replication, but probably only in small numbers and only in special circumstances (Golde *et al.*, 1974a,b).

In Collaboration with Albert D. Diesseroth and Jacqueline Whang-Peng from the National Cancer Institute, we have followed the replacement of host alveolar macrophages in a male patient who received a successful bone marrow transplant from his sister (Fig. 6). Alveolar macrophages of male (host) origin were identified by the presence of a fluorescent Y chromosome. Although the bone marrow and peripheral blood cells of this patient were all of female (donor) origin, significant proportions of his alveolar macrophages were of male (host) origin for at least 4.5 years posttransplantation. These studies suggest that, at least in this circumstance, alveolar macrophages are capable of replication *in situ* or, alternatively, that the half-life of these cells is very long.

IV.   Perspectives

Human alveolar macrophages play a major role in the defense of the lung in health and in disease states. In this capacity, these cells remove debris from the alveolar surface and are the first line of defense against most microorganisms that invade the lower respiratory tract. In addition, as a result of these activities, it is likely that alveolar macrophages prevent most antigens from gaining access to other arms of the immune system, thereby sheltering the lung and other organs from continual inflammatory and immune-mediated damage.

It is clear, however, that the alveolar macrophage can function in many capacities other than its traditional role as the premier phagocyte of the lung. As a result of these other functions, it is now clear that the alveolar macrophage can injure the lung as well as protect it. Recent studies demonstrating that alveolar macrophages of smokers may spontaneously secrete elastase are important in this regard (Rodriguez *et al.*, 1977). In addition, alveolar macrophages can also injure the lung by secreting various chemotactic factors which attract neutrophils to the lung (Kazmierowski *et al.*, 1977; Hunninghake *et al.*, 1978). These neutrophils can also injure the lung parenchyma by virtue of their large stores of lysosomal proteases (Gadek *et al.*, 1979). Alveolar macrophages may also be important in the pathogenesis of lung cancer by detoxifying carcinogenic compounds (Cantrell *et al.*, 1973) or by activating precarcinogens in inhaled particulates such as those present in cigarette smoke (Harris *et al.*, 1978).

Studies of the human alveolar macrophage are, for the most part, still in their infancy. However, with the recent ready availability of these cells through the use of fiberoptic bronchoscopy, it is likely that these studies will be expanded considerably in the near future.

REFERENCES

Acton, J. D., and Myrvik, Q. N. (1966). *J. Bacteriol.* **91,** 2300–2304.

Blondin, J., Rosenberg, R., and Janoff, A. (1972). *Am. Rev. Respir. Dis.* **106,** 477–479.

Brody, A. R., and Craighead, J. E. (1975). *Lab. Invest.* **32,** 125–132.

Campbell, E. J., White, R. R., Senior, R. M., Rodriguez, R. J., and Kuhn, C. (1979). *J. Clin. Invest.* **64,** 824–833.

Cantrell, E. T., Warr, G. A., and Martin, R. R. (1973). *J. Clin. Invest.* **52,** 1881–1884.

Cohen, A. B. (1973). *J. Clin. Invest.* **52,** 2793–2799.

Cohen, A. B., and Cline, M. J. (1971). *J. Clin. Invest.* **50,** 1390–1398.

Condon, W. L., and Harris, H. O. (1978). *Chest* **73,** 364–370.

Crystal, R. G., Fulmer, J. D., Roberts, W. C., Moss, M. L., Line, B. R., and Reynolds, H. Y. (1976). *Ann. Intern. Med.* **85,** 769–788.

Drew, W. L., Mintz, L., Hoo, R., and Finley, T. N. (1979). *Am. Rev. Respir. Dis.* **119,** 287–291.

Finley, T. N., Swenson, E. W., Curran, W. S., Huber, G. L., and Ladman, A. J. (1967). *Ann. Intern. Med.* **66,** 651–658.

Gadek, J., Fells, G., Hunninghake, G., Zimmerman, R., and Crystal, R. G. (1979). *Clin. Res.* **27,** 397A.

Golde, D. W., Finley, T. N., and Cline, M. J. (1974a). *N. Engl. J. Med.* **290,** 875–878.

Golde, D. W., Byers, L. A., and Finley, T. N. (1974b). *Nature (London)* **247,** 373–375.

Golde, D. W., Territo, M., Finley, T. N., and Cline, M. J. (1976). *Ann. Intern. Med.* **85,** 304–309.

Goldstein, E., Lippert, W., and Warshaver, D. (1974). *J. Clin. Invest.* **54,** 519–528.

Green, G. M. (1970). *Am. Rev. Respir. Dis.* **102,** 691–703.

Green, G. M., and Kass, E. H. (1964). *J. Exp. Med.* **119,** 167–176.

Harris, C. C., Hsu, I. C., Stoner, G. D., Trump, B. F., and Selkirk, J. K. (1978). *Nature (London)* **272,** 633–634.

Harris, J. O., and Goldblatt, J. (1976). *Clin. Res.* **24,** 385A.

Harris, J. O., Swenson, E. W., and Johnson, J. E. (1970). *J. Clin. Invest.* **49,** 2086–2096.

Harris, J. O., Olsen, G. N., Castle, J. R., and Maleney, A. S. (1975). *Am. Rev. Respir. Dis.* **111,** 579–586.

Horwitz, A. L., and Crystal, R. G. (1976). *Biochem. Biophys. Res. Commun.* **69,** 296–303.

Horwitz, A. L., Kelman, J. A., and Crystal, R. G. (1976). *Nature (London)* **264,** 772–774.

Hunninghake, G. W., and Fauci, A. S. (1976a). *Cell. Immunol.* **26,** 89–97.

Hunninghake, G. W., and Fauci, A. S. (1976b). *Cell. Immunol.* **26,** 98–104.

Hunninghake, G. W., Gadek, J. E., Kawanami, O., Ferrans, V. J., and Crystal, R. G. (1979). *Am. J. Pathol.* **97,** 149–206.

Hunninghake, G. W., Gallin, J. I., and Fauci, A. S. (1978). *Am. Rev. Respir. Dis.* **117,** 15–23.

Kazmierowski, J. A., Gallin, J. I., and Reynolds, H. Y. (1977). *J. Clin. Invest.* **59,** 273–281.

Langevoort, H. C., Cohn, Z. A., Hirsch, J. G., Humphrey, J. H., Spector, W. G., and van Furth, R. (1970). *In* "Mononuclear Phagocytes" (R. van Furth, ed.), Vol. I, pp. 1–6. Davis, Philadelphia, Pennsylvania.

Laughter, A. H., Martin, R. R., and Twomey, J. J. (1977). *J. Lab. Clin. Med.* **89,** 1326–1332.

Leffingwell, C. M., and Low, R. B. (1975). *Am. Rev. Hosp. Dis.* **112,** 349–359.

Levine, E. A., Senior, R. M., and Butler, J. V. (1976). *Am. Rev. Respir. Dis.* **113,** 25–30.

Liu, C. (1955). *J. Exp. Med.* **101**, 665–676.

Lussier, L. M., Chandler, D. K. F., Sybert, A., and Yeager, H. (1978). *J. Appl. Physiol.* **45**, 933–938.

Myrvik, Q. N., and Acton, J. D. (1977). *In* "Respiratory Defense Mechanisms" (J. D. Brain, D. F. Proctor, and L. M. Reed, eds.), Vol. V, Part II, pp. 1023–1051. Dekker, New York.

Nichols, B. A. (1976). *J. Exp. Med.* **144**, 906–919.

Olsen, G. N., Harris, J. O., Castle, J. R., Waldman, R. H., and Karmgard, H. J. (1975). *J. Clin. Invest.* **55**, 427–430.

Reynolds, H. Y., and Newball, H. H. (1974). *J. Lab. Clin. Med.* **84**, 559–573.

Reynolds, H. Y., Atkinson, J. P., Newball, H. H., and Frank, M. M. (1975). *J. Immunol.* **114**, 1813–1819.

Reynolds, H. Y., Fulmer, J. D., Kazmierowski, J. A., Roberts, W. C., Frank, M. M., and Crystal, R. G. (1977). *J. Clin. Invest.* **59**, 165–175.

Rodriguez, R. L., White, R. R., Senior, R. M., and Levine, E. A. (1977). *Science* **198**, 313–314.

Schuyler, M. R., Thigpen, T. P., and Salvaggio, J. E. (1978). *Ann. Intern. Med.* **88**, 355–358.

Simon, L. M., Robin, E. D., Phillips, J. R., Acevedo, J., Apline, S. G., and Theodore, J. (1977). *J. Clin. Invest.* **59**, 443–448.

Thomas, E. D., Ramberg, R. E., Sale, G. E., Sparkes, R. S., and Golde, D. W. (1976). *Science* **192**, 1016–1018.

Ulrich, F. (1977). *Res. J. Reticuloendothel. Soc.* **21**, 33–51.

Unanue, E. R. (1976). *Am. J. Pathol.* **83**, 396–417.

Warr, G. A., and Martin, R. R. (1973). *Am. Rev. Respir. Dis.* **108**, 371–373.

Warr, G. A., and Martin, R. R. (1974). *Infect. Immun.* **9**, 769–771.

Weinberger, S. E., Kelman, J. A., Elson, N. A., Young, R. C., Reynolds, H. Y., Fulmer, J. D., and Crystal, R. G. (1978). *Ann. Intern. Med.* **59**, 459–466.

Zwilling, B. S., and Campolito, L. B. (1977). *J. Immunol.* **119**, 838–841.

Chapter 7

Culture of Cells and Tissues from Human Lung—An Overview

CURTIS C. HARRIS

Human Tissue Studies Section,
Laboratory of Experimental Pathology,
National Cancer Institute,
Bethesda, Maryland

The lung is one of the most complex human organs. Approximately 40 different cell types have been identified in the mammalian lung. A second striking anatomical feature is the several hundred square meters of surface area of respiratory epithelium in this organ. While essential for gaseous exchange of oxygen and carbon dioxide, this extensive surface area is also exposed to inhaled environmental pollutants. These pollutants may both directly damage the lung and, following absorption, cause systemic injury. Pathological changes in the lung have been described in detail. In contrast, the biochemical and physiological functions of the lung are less well known. Model systems with cultured human tissues and cells can be used to investigate normal functions of the lung as well as pathological alterations in this vital organ. The ability to maintain human tissues and cells in a controlled experimental setting is an important facet of such model systems.

The first six chapters of this volume provide the reader with the current status of methodology for culturing explants and cells from human lung. In Chapter 1, Trump *et al.* describe methodology recently developed for the culture of human and bovine bronchial explants for time periods of months. Bronchial specimens are collected at the time of surgery and at immediate autopsy. A remarkable finding is that these human bronchial specimens are viable after storage in cold (4°C) L-15 tissue culture medium for several days prior to culture. This minimizes logistical problems associated with transporting tissue to the laboratory and makes possible the development of collaborative studies between laboratories. The culture conditions are still being defined. For example, "enriched" culture media that stimulate cell proliferation and perhaps epithelial

113

repair of "wounding" may not maintain normal mucociliary epithelium in bronchial explants for time periods as long as a more simple medium such as minimal Eagle's medium. Depending on the purpose of the study, one may select culture conditions that either maintain normal cell differentiation or enhance cell proliferation.

While methods for culturing bronchial explants have been available for several years, monolayers of human bronchial epithelial cells have been more difficult to maintain in culture. In Chapter 2, Stoner *et al.* review previous attempts to culture these cells and describe the current status of their investigations. Their approach has been initially to utilize culture conditions that enhance outgrowth of epithelial cells from the bronchial explants. Methods for quantitatively measuring epithelial outgrowth have been developed so that the culture conditions, including medium composition, substrate, atmosphere, etc., can be refined. In addition, precise identification of the major types of bronchial cells by morphological, cytochemical, immunochemical, and biochemical methods are described. The presence of the explant enhanced the period of time the epithelial outgrowth could be maintained. In the epithelial outgrowth, the normal mucociliary epithelium was replaced by a squamous keratinizing epithelium. The epithelial cell monolayers could be transferred for three to four passages before cessation of cell proliferation. While the optimal culture conditions for modulating growth and/or normal differentiation have not as yet been developed, cultured human bronchial epithelial cells are being utilized in studies of cell biology and pathology.

The simple morphology of fibroblast cells has been equated with simple function. Bradley *et al.* (Chapter 3) dispel this incorrect assumption by describing several morphological phenotypes of fibroblasts and their varied and complex biochemical functions. Since methods for culturing human fibroblasts have been kown for many years, they describe the major approaches used to culture this cell type. More than 20 fibroblast cell strains have been cultured from human lung. Since one of the functions of the fibroblast is to aid in maintaining alveolar structure, collagen types I and III are biochemical markers of these cells and their synthesis has been extensively studied.

When compared to bronchial explants the long-term culture of peripheral human lung has been less successful, and the normal architecture of the tissue is generally lost after a few days in culture. The optimal culture conditions currently available are described by Stoner in Chapter 4. Fragments of adult lung are placed on pieces of gelatin sponge in tissue culture dishes, and the cultures are placed in a rocking chamber to permit intermittent exposure of the fragments to the atmosphere and to the culture medium. During the 25 days in culture, type-2 alveolar epithelial cells proliferated and gradually replaced the type-I cells, such that the alveolar structures were composed entirely of type-2 cells. This short-term culture system has been used to study the metabolism of a carcinogenic polynuclear aromatic hydrocarbon and may be useful for studies on the biosynthesis of surfactant and collagen.

An organotypic culture system for the maintenance of human and rat type-2 alveolar cells is described by Douglas *et al.* (Chapter 5). By dissociating and then reaggregating fetal lung, organotypic cultures composed predominantly of type-2 cells are formed. The alveolar structure is maintained for several days so that the maturation of these cells and the synthesis of surfactant phospholipids can be investigated.

Alveolar macrophages phagocytize inhaled particulates, including microbial agents, from alveolar surfaces. Methods for culturing these cells, as well as for studying their multiple functions, are discussed by Hunninghake *et al.* in Chapter 6. Macrophages collected via either bronchoalveolar lavage or lung biopsy can be separated from other lung cells by density gradient centrifugation and by differential attachment of the macrophages to a solid substrate. Human alveolar macrophages can be maintained in monolayer culture for at least 60 days. These cultured macrophages, (1) ingest and kill microorganisms, (2) secrete a variety of enzymes including proteases, collagenase, and elastase, (3) secrete soluble immune mediators, and (4) may be accessory cells in immune reactions.

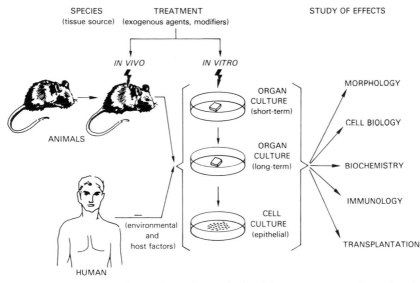

FIG. 1. Comparative studies on tissues from animal and human sources can be conducted by culture methods. Exogenous agents, hormones, and modifiers can be administered *in vivo* in animal models and *in vitro* in both animal and human systems. Different combinations of experimental treatment can therefore be designed. Cell cultures can be derived from previously established organ cultures. Each level of organization can be analyzed concurrently by morphological, immunological, and biochemical techniques and by observation of cellular properties and growth characteristics in culture as well as after transplantation into immunocompatible animals.

If the progress in methods for culturing cells and tissues from human lung continues to accelerate at the current rate, the remaining problems will soon be solved. For example, the development of chemically defined media for the long-term culture of monolayers of lung cells is a major challenge that will be met. Concomitant with efforts to improve methodology, progress will also be made in many areas of biomedical research by using currently available culture methods. Model systems with cultured human tissues and cells are being used to study the pathogenesis of human diseases.

The general approach for these studies is illustrated in Fig. 1. In the area of chemical carcinogenesis, the metabolism of chemical carcinogens is being investigated in cultured human bronchus, lung, and macrophages. Model systems can be developed for studying the pathogenesis of pulmonary fibrosis and associated abnormalities in collagen synthesis. The differentiation and maturation of human fetal lung is another area of study that can be conducted with these model systems. As shown in Fig. 1, the parallel investigation of tissues and cells from experimental animals is an important feature of this model system approach. Such comparative studies between tissues from humans and experimental animals offer exciting opportunities to aid our understanding of the normal biology of the lung and the pathogenesis of diseases in this vital organ.

Chapter 8

Culture of Aorta

H. PAUL EHRLICH[1]

Shriner Burns Institute,
Massachusetts General Hospital,
Harvard Medical School,
Boston, Massachusetts

I. Introduction

Development of Organ Culture Technique

This technique, as a method for studying physiological and pathological events in a controlled environment, was originated at Strangeways Research Laboratories in England more than 50 years ago (Strangeways and Fell, 1926).

---

[1]This work was done while the author was a research fellow of the American and British Heart Associations at the Tissue Physiology Department, Strangeways Research Laboratory, Cambridge, England.

Fell and Robison (1929) grew a developing embryonic limb bud in culture. These limbs grew and differentiated in a controlled *in vitro* environment. Embryonic tissues do well in culture because of their ability to grow under ischemic conditions. Mature tissues usually have difficulty in organ culture because of the ischemic conditions. The tissue matrix of mature explants in cultures can become altered, and fibroblast-like cells will proliferate from these explants and grow out as a monolayer on a glass or plastic surface.

1. BLOOD VESSELS IN ORGAN CULTURE

Trowell (1956) examined a number of organs in culture and concluded that explants that were naturally thin or flat, or that flattened out in culture, generally did better than explants that remained in a spherical or cubical shape. He placed ring segments of rat arteries on lens paper which was placed on stainless steel grids sitting in a defined culture medium under a moist atmosphere of air and carbon dioxide. After 9 days in culture, the artery explants were unchanged. The histological appearance of this tissue was perfectly preserved, with healthy endothelium, media, and adventitia.

Since a major part of the medial wall of the blood vessel survives *in vitro* by diffusion of nutrients from the circulating blood, its chances of survival in organ culture would be expected to be excellent. Fritz *et al.* (1976) maintained 1.2-mm squares of swine aorta medial wall as a free-floating explant in organ culture for 9 days in M199 medium with additions of human sera. The protein synthetic activity of the 9-day explant was equivalent to that of freshly extracted tissue.

Blacklow *et al.* (1975) used samples of human fetal aorta in Leibovitz L-15 medium, with additions of 0.2% bovine albumin, glutamine, and antibiotics, and the explants maintained their integrity for 8 weeks. These explants were ring segments of vessel placed directly on petri dish surfaces with the medium only partially covering them. Histologically, their endothelium was in perfect health after 8 weeks in culture. In another study, St. Clair and Harpold (1975) maintained rectangular segments of pigeon aorta with the endothelial surface placed down on stainless steel mesh grids for 9 days. The explants were cultured in Dulbecco's modification of Eagle's medium (DMEM), with the addition of 5–60% pigeon serum. The serum was taken from pigeons maintained on a normal diet or on a high-cholesterol diet, and the effects of these sera on cholesterol esterification were measured.

2. AORTA IN STATIONARY AND ROLLER CULTURE

We report here that stationary cultures of ringed segments of normal or injured rabbit aorta can be maintained in organ culture for periods up to 11 days. Larger segments of vessels (1.5 cm long) can be maintained in a roller culture for 12 days. These culture systems were developed in order to study biochemical and histological aspects of experimental atherosclerotic-like lesions in vessel under a controlled environment. The cultures were useful for identifying the cell population that secretes the lysosomal enzyme, cathepsin D, using an antibody-trapping technique developed at Strangeways Research Laboratories (Poole *et al.*, 1976). New collagen synthesis was monitored in a controlled environment using the roller culture technique.

II. Methods and Materials

A. Rabbit Aorta

New Zealand White adult male rabbits were divided into treatment groups of six animals each. Group I included control rabbits maintained on a normal diet of commercial rabbit chow. Group II consisted of injury-alone rabbits that had intimal injury to their aortas and were maintained on a normal diet of commercial rabbit chow. Group III included injury-with-cholesterol-diet rabbits which were maintained on a diet of rabbit chow supplemented with cholesterol (1% by weight). After 3 months on this diet, the rabbits' cholesterol serum ranged from 700 to 1100 mg %. The injury to rabbits in groups II and III was the scraping and removal of the endothelial surface of the thoracic and abdominal aortas (Stemerman and Ross, 1972). Rabbits were placed under a light anesthetic using pentobarbitol. The groin area was shaved and washed with 70% alcohol. Clean surgical technique was employed in making an incision, and the femoral artery was exposed using blunt dissection. A Fogarty F4 balloon catheter was introduced through a small incision in the femoral artery and pushed up through the abdominal and thoracic aortas to the aortic bifurcation. The balloon catheter was inflated and pulled back to the femoral artery. The balloon was deflated, and the catheter removed from the femoral artery which was tied off in order to stop any bleeding. The lateral circulation was more than adequate for maintaining a blood supply to the rabbit's hind limb. The wound was washed out with sterile saline, 200 mg of topical penicillin was sprinkled on it, and a single injection of 100 mg of penicillin was injected intramuscularly. The wounds were sutured closed, and the animals returned to their cages and placed on their appropriate diets.

B. Preparation of Aorta Explants

1. STATIONARY ORGAN CULTURE

In the experiments described rabbits were maintained on their diets for 3 months following ballooning. At this time they were killed by cervical dislocation, and the abdomen opened. Sixty milliliters of blood was drawn from the exposed vena cava and allowed to clot in a glass container. The blood serum was separated from the blood clot, sterilized by filtration, heat-treated for 30 minutes at 56°C, and stored at −20°C until needed for supplementing culture medium. After the bleeding, the thoracic cavity was opened, and both the thoracic and abdominal aortas were exposed by blunt dissection and removed. The aortas were placed in DMEM which included 1000 U of penicillin and 1000 μg of streptomycin per milliliter. They were taken to a sterile flow hood, and the excess adventitia removed by careful dissection. The aortas were washed at least four times during this washing period with additional fresh DMEM containing antibiotics. All these steps were carried out at room temperature. It was noted that segments of aorta could be left in DMEM with antibiotics overnight at 4°C without impairing their viability in culture.

For stationary organ cultures, ring segments 2–3 mm long were cut (Fig. 1A). All instruments were briefly swirled in boiling water before use and laid down on a sterile surface between uses. The small ring segments were placed with the cut edge down on a Millipore filter (0.45-μm pore size) which was placed on top of a 1.2 × 1.2 cm stainless steel grid bench with two parallel edges folded down at a depth of 0.5 cm (Fig. 1B). Medium (2.5 ml) was added to a 35 × 15 mm petri dish. This volume was just enough to cover the surface of the stainless steel grid. The rings of aorta, usually four rings per dish, were not covered with medium. The petri dishes were placed in a 5% carbon dioxide −95% air atmosphere. The incubation medium was changed every other day.

The medium used for these cultures was DMEM with 100 U penicillin per milligram, 50 μg of streptomycin per milliliter, and serum added to a final volume of 10%. The serum for these cultures was either rabbit serum from rabbits on a normal diet or rabbits maintained on a 1% cholesterol diet. The serum was heat-treated at 56°C for 30 minutes, sterilized by filtration, and stored at −20°C until needed. Aorta could be maintained for 2 days in serum-free medium without histological damage.

2.   ROLLER ORGAN CULTURE

For roller cultures the processing of the aorta was exactly the same as for stationary cultures. Aorta segments 1.5 cm in length were used (Fig. 1A). A

FIG. 1.   Preparation of stationery and roller organ cultures. (A) Rabbit thoracic aorta cut into segments. On the right is a ring (cross view) used in stationery organ culture. On the right is a single segment of vessel (side view) used in roller organ culture. (B) Stationary organ culture in a petri dish, showing two aortic ring segments on a Millipore filter on top of a stainless steel mesh grid submerged in DMEM. (C) Components of the roller organ culture. At the lower right is a 15-mm segment of vessel sutured onto a stainless steel mesh disk. Above it is a stainless steel spring which holds the disk at the bottom of the Universal vial shown on the left. The plastic screw cap for the vial is shown above the spring. The assembled roller organ culture is shown with an explant on the disk which is held in place within the Universal vial by the spring.

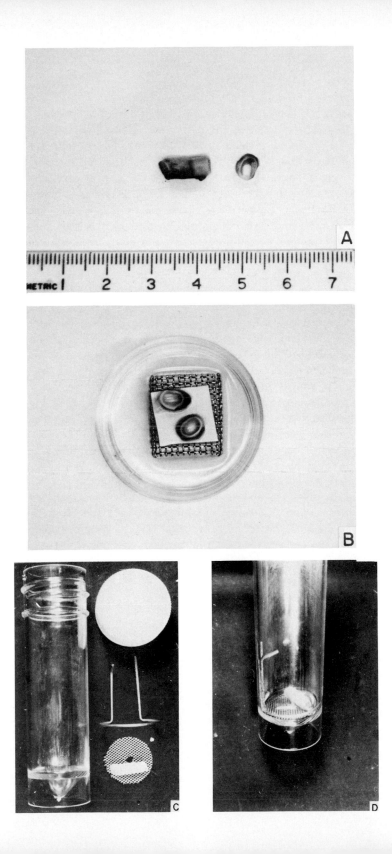

stainless steel disk 3 cm in diameter was held in a bulldog clamp which in turn was clamped in a Universal joint vise (Silvo Hardware, Philadelphia). The bulldog clip, vise, and stainless steel mesh disk were cleaned with 70% ethanol and allowed to air-dry in the laminar flow. One aorta segment was placed on the periphery of the stainless steel disk and carefully balanced there. A 23-gauge sterile hypodermic needle was pushed through the aorta segment and into and through the screen mesh underneath it. A piece of 3-0 nylon suture was pushed through the needle, and the needle was drawn back through the mesh disk and aorta, leaving the suture material protruding through the aorta segment. The suture material was rethreaded through the mesh disk, and a surgical knot was tied on the underside of the disk with sterile forceps. When both ends of the aorta segment had been attached to the disk in this manner, the disk was transferred, segment side up, to a Universal vial (York Scientific, Inc., Ogdenburg, New Jersey) (Fig. 1C).

The aorta segment–disk assembly was held in place at the bottom of the Universal vial by a spring arrangement made of stainless steel (Fig. 1D). Three milliliters of DMEM with 10% rabbit serum was added to each vial, and the screw cap attached loosely. The cap was held in place by the addition of a piece of tape. This permitted a free exchange of gases between the vial and its surrounding environment in the incubator. The vial with its aortic segment was placed in a Bellco roller drum (Bellco Glass, Inc., Vineland, New Jersey) having thirty 2.7-cm-diameter holes in it and placed on a Bellco roller rack set to rotate at 1 rpm. The roller rack assembly was placed in a moist atmosphere of 95% air–5% carbon dioxide at 37°C. The roller rack was tilted at an angle of about 75°C in order to keep medium at the proper level within the vial. As the roller rack rotated, the aorta was first submerged in the medium and then removed from it so that the medium could circulate constantly over and through the vessel explant. The medium was changed every other day by simply pouring off the used medium and replacing it with 3 ml of fresh medium.

C.   Analysis

1.   Histology

The stationary cultures were maintained from 3 to 11 days, and the segmented aortic rings then fixed in 10% Formalin and processed for light microscope studies. In order to test for viability of the explants, 1 μCi/ml of [^{3}H]proline was added to the cultures 24 hours prior to harvesting. The rings were removed, fixed, and processed for autoradiography. They were again examined by light microscopy.

Further histological studies were done using an electron microscope. Some aorta segments were fixed, stained, and examined by a Philips 300 transmission electron microscope in the Pathology Department, Cambridge University.

2. CATHEPSIN D

Cathepsin D is a lysosomal acid protease which has been shown to be elevated in atherosclerotic vessels (Peters *et al.*, 1972). Biochemical analysis of cathepsin D in these aortic explants and in fresh aorta was made by homogenizing in 150 mM sodium chloride solution with 0.1% Triton X-100. The homogenate was cleared by centrifugation, and the supernatant assayed for cathepsin D activity using the method of Barrett (1972). The concentration of cathepsin D was measured by the number of units of cathepsin D per milligram wet weight of tissue and per milligram of protein, using a modified method of Lowry *et al.* (1951).

Localization of cathepsin D secretion within the vessel was accomplished using the method developed by Poole *et al.* (1974). Stationary organ cultures were incubated for 2 days with heat-treated sterile normal sheep serum or heat-treated sheep antiserum to rabbit cathepsin D (Weston, 1969). Heat treatment was 30 minutes at 56°C. The medium was changed, and the explants washed with serum-free medium two times and then incubated for an additional 24 hours in serum-free DMEM. In other experiments, the F(ab)$_2$ fraction from IgG of normal and immunized sheep sera was purified by 50% ammonium sulfate precipitation followed by 48 hours of incubation at 37°C in 0.5 M acetic acid with 1% w/v pepsin and the pH adjusted to 3 with hydrochloric acid. The F(ab)$_2$ fragment was purified by gel filtration and then added to cultures and incubated in 10% rabbit serum at a concentration of 5 mg/ml. This treatment of IgG was carried out to remove any effect sheep sera might have on vessels in culture and to eliminate any nonspecific binding between sheep IgG and rabbit aorta. After the incubation with antibody and the washing out of unbound F(ab)$_2$, the explants were placed in a flat-bottomed plastic tube, 1 ml of 7% warm gelatin solution in phosphate-buffered saline (PBS) was added, and the tube immediately submerged in liquid nitrogen for 90 seconds. The frozen explants were sealed and kept at −20°C until needed. The plastic tube was then cracked, and the frozen gelatin-embedded specimen removed and sections cut with a cyrostat. Sections approximately 7 μm thick were cut and placed on glass slides. All the subsequent procedures were carried out at room temperature. Sections were fixed in freshly prepared 4% paraformaldehyde buffered in PBS with the pH adjusted to 7.4. After a 2-minute fixation period, the slides were washed four times for 10-minute periods in PBS. The sheep IgG antibody–antigen complex in aorta explants was localized by incubating cut sections for 30 minutes with rabbit (anti-sheep IgG) F(ab)′ fragment labeled with fluorescein isothiocyanate (FITC) following the methods of Poole *et al.* (1974). The unbound rabbit (anti-sheep IgG) F(ab)′ FITC was removed from the sections by washing four times in PBS with 5 mM cysteine for

15-minute periods. The sections were then counterstained with eriochrome black and mounted in a mixture of 0.2 M Tris–hydrochloride buffer, pH 8.6, with glycerol (1:9 v/v) (Poole et al., 1974).

The sections were examined with a Reichert Zetopan microscope using dark-field illumination. The specimens were excited with light passing through a number 104C filter (cutoff 500 nm) and a BG 38 filter. The emission was photographed through a OG 527 barrier filter. All filters were from Polaron Equipment, Ltd., London, England. Following fluorescence examination, the coverslip was removed and the slide washed with water, fixed in 10% formaldehyde, stained with hematoxylin and eosin, and reexamined by light microscopy.

To summarize this method of localizing extracellular cathepsin D as an antibody–antigen complex, first, specimens of aorta control, injured vessel–normal diet, and injured vessel–cholesterol diet were incubated in normal sheep serum or sheep antiserum to rabbit cathepsin D. In other experiments the F(ab)$_2$ of normal and immunized sheep sera IgG were incubated with different specimens of rabbit aorta incubated with rabbit serum. The organ cultures were incubated for 24 hours in serum-free medium to allow free sheep IgG to diffuse out of the tissues and leave behind any insoluble immunocomplex. The complexes were detected by cutting frozen sections of organ-cultured aorta and localizing sheep IgG with fluorescence microscopy using fluorescein-labeled rabbit antibody to sheep IgG.

3. COLLAGEN SYNTHESIS

The monitoring of collagen synthesis was accomplished with roller cultures. Proline (2μCi/ml) was incubated with the explants for 24 hours in the absence of serum but in the presence of 10μg/ml of β-aminoproprionitrile, which prevents collagen cross-linking, and 50μg/ml of sodium ascorbate. The explants were homogenized in 0.5 M acetic acid with 0.1 mg/ml pepsin in order to convert any collagen precursors to adult collagen forms. Pepsin digestion also removed non-collagen protein. The digestion was carried out at 4°C for 24 hours with stirring, and this time the homogenate was cleared by centrifugation. The pepsinized insoluble digest was discarded. The pepsin was inactivated by adjusting the pH of the supernatant to 8.0 with 1 M sodium hydroxide, and dry sodium chloride was added to a final concentration of 20% w/v. The salted supernatant mixture was stirred at 4°C overnight and then cleared by centrifugation at 39,000 g for 10 minutes. The labeled collagen-enriched pellet was dialyzed exhaustively versus 0.1 M acetic acid and lyophilized to dryness. A fraction of this dry material was subjected to 6% sodium dodecyl sulfate polyacrylmide gel electrophoresis (SDS-PAGE). The gels were stained, destained, and then washed in water. The gels were impregnated with PPO, and fluorograms were developed according to the method of Bonner and Laskey (1974). Electrophoretic standards for type I collagen were run for comparison.

III. Results

A. Morphological Findings

The gross appearance of the explants in both the stationary and roller cultures was unremarkable after 12 days of culture. The medium was noticeably more acidic in some roller cultures than in the stationary cultures after 48 hours of culturing. However, no significant histological changes were noted when the medium in the roller cultures was changed every day. So far there have been no contamination problems involving the roller cultures, but about 5% of the stationary cultures had become contaminated.

1. HISTOLOGY

The histological appearance of the stationary cultures showed a well-preserved matrix and vessel wall in 90% of cultures. The atherosclerotic-like lesions in the vessel intima were unchanged during this culturing period. In some areas there was slight swelling in the medial walls at 7–12 days. But for the most part, the overall structure was well preserved. Cell viability was usually excellent in the adventitia during these studies (Fig. 2A). Any problem involving cellular death occurred in the medial wall. In 70% of the cultures maintained for more than 10 days, the medial wall, intima, and adventitia were viable. In the other 30%, large areas of the medial wall were acellular, but the intima and adventitia were in good condition. In three of the 72 specimens examined there was a new growth of cells in the lumen area of the explant (Fig. 2B). In one case, serial sectioning revealed that the cells in the adventitia had grown through an intercostal aorta opening and into the center of the vessel explant. In two other cases, cells in the adventitia had grown over the top cut edge of the ring and about ½ mm down into the lumen. The use of larger segments in the roller culture eliminated this problem, and so far this ingrowth of adventitia cells has not appeared in over 60 specimens examined. The electron microscope examination, made by Malcolm Mitchison, Department of Pathology, University of Cambridge, Tennis Court Road, Cambridge, showed residual bodies within cells of the atheroma-like lesion (Fig. 2C). The extracellular matrices of the vessels were in satisfactory condition after 1 week in culture. Smooth muscle connective tissue cells present had mitochondria, rough endoplasmic reticulum, and a Golgi complex. In general, normal vessels and atheromatous vessels did well in stationary cultures morphologically

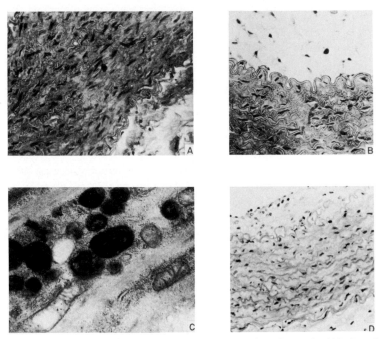

FIG. 2.   Histology of aorta in organ culture. (A) Cross section of normal rabbit thoracic aorta after 6 days in a stationary organ culture. This explant was maintained in DMEM supplemented with 10% normal rabbit serum. The adventitia of the vessel wall is at the lower right, and the lumen of the aorta at the upper right. The vessel wall appears to be in good condition. (B) Normal rabbit abdominal aorta after 11 days in stationary organ culture. This explant was maintained on DMEM with 10% normal rabbit serum. The upper half of the figure shows a loose connective tissue matrix filling the lumen of the vessel segment. This matrix is formed by the fibroblasts of the adventitia growing over the top edge of the explant and moving down the intimal wall. (C) Electron micrograph of an injured vessel–cholesterol diet rabbit abdominal aorta explant after 6 days in stationary organ culture. The area being examined is the cytoplasm of a cell within the intimal thickening of the aorta. Mitochondria and residual bodies are seen in this section. (D) Cross section of scraped abdominal aorta from a rabbit on a cholesterol diet after 7 days in roller organ culture. The rabbit was injured and maintained on the cholesterol diet for only 5 days prior to explanting the aorta. The vessel was cultured in DMEM with 10% cholesterol-diet rabbit serum. The lumen of the aorta is at the upper left, where a thickened intima has developed in this short 12-day period, i.e., 5 days *in vivo* and 7 days *in vitro*.

and histologically. In these explants no outgrowth of the medial wall was observed in any of the specimens studied. This is attributed to the Millipore filter surface in the stationary cultures and the stainless steel grid surface in the roller cultures.

In one experiment a rabbit fed a normal diet was killed 5 days after ballooning. The vessel was placed in a roller culture with 10% normal rabbit serum for 7 days. Histological examination of the vessel showed an impressive growth in the intimal space (Fig. 2D), and a great deal of this intimal growth must have occurred in culture. Numerous attempts have been made to repeat these results, but the amount of intimal growth obtained so far has not been as great as that first seen, although some growth has been seen in all aorta explants. Injury to the intima or medial wall while the vessel is in organ culture will result in cellular death and necrosis. Sade *et al.* (1972) reported that injury *in vivo* 24–48 hours before explanting aorta resulted in the simulation of endothelial cell mitosis. In our experiments, however, injury occurring *in vitro* resulted in cellular death and no apparent regeneration.

2.   Autoradiographic Results

Autoradiography studies of stationary cultures revealed that grains were visible over cells surrounding the aorta in cross section. The adventitia showed the highest density of grains over cells within the loose connective tissue of this area (Fig. 3A), and the medial wall showed an even density of grains over cells throughout this area (Fig. 3B). This was true in both roller cultures and stationary cultures. In the endothelial wall of normal vessels there was high activity associated with the endothelial cells on the internal elastica lamina (Fig. 3B). In the atheromatous intima the highest density of grains was actually over the cells on the lumenal side of the lesion (Fig. 3C). Cells closest to the medial wall had the least amount of activity. The density of the grains over these cells closest to the medial wall was the lowest of any of the cells examined in the whole section. This indicated that most active cells in the atheromatous-like plaque were those in the lumenal area, and that the least active synthetic cells in the whole preparation were those in the lesion closest to the medial wall.

The roller culture explants showed uniform labeling of the medial wall. In general, aorta explants were synthetically active. Cells in the adventitia were labeled quite heavily, much more so than those in the medial wall. The cells in the intimal space and the atheromatous-like plaque were labeled differently. In the intimal lesion, the cells closest to the medial wall were labeled less than the cells closest to the lumen. Since there was uniform labeling throughout the medial wall, diffusion would not be expected to be responsible for the difference in cellular synthetic activity of the cells within atheromatous-like lesions. It is concluded that in the atheromatous-like lesion, produced by ballooning and cholesterol feeding, there are two cell populations, one with high synthetic activity and the other with mdoerate or low synthetic activity.

B.   Biochemistry

1.   Cathepsin D

The concentration of cathepsin D per milligram wet weight from vessel was 5.8 mU in the normal diet–uninjured vessel group, 5.1 mU in the injured

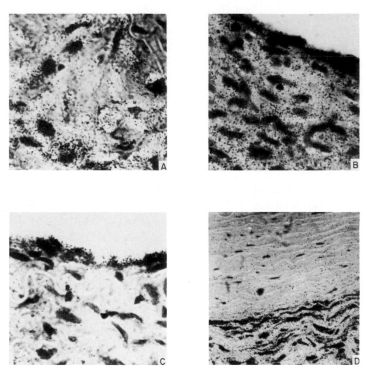

FIG. 3.   Autoradiographs of aorta in organ culture pulse-labeled for 24 hours with [³H]proline.
(A) A high-power field of the adventitia from a normal rabbit aorta after 7 days in stationary organ
culture. The explant was maintained in DMEM with 10% normal rabbit serum. On day 6 the medium
was changed, with 2.5 μCi [³H]proline added, and the explant was incubated for another 24 hours.
The external elastica lamina is at the top right. Silver grains are prominent over the fibroblasts within
the adventitia layer. (B) Cross section of a normal rabbit aorta after 7 days in roller organ culture. The
explant was incubated for 24 hours with 3 μ Ci [³H]proline in 3 ml DMEM with 10% normal
rabbit serum. The lumen of the aorta is in the top right corner. The highest density of silver grains is
over the endothelial cells lining the vessel wall. There is an even density of silver grains over the
smooth muscle cells of the medial wall. (C) Section through an intimal thickening of a rabbit
abdominal aorta. The rabbit was maintained on a 1% cholesterol diet, and the endothelial lining of the
vessel had been scraped off 3 months prior to explanting in a stationary organ culture. The explant
was kept for 7 days in DMEM with 10% normal rabbit serum. It was pulsed-labeled with 2.5 μCi
[³H]proline 24 hours prior to harvesting. Within thickened intima, there is a high density of silver
grains over the cells nearest the lumen of the aorta, and a lower density over the cells nearest the
internal elastica lamina. (D) A cross section at lower power of a human mesentary artery. The artery
was kept for 9 days in roller organ culture with DMEM and 10% fetal bovine serum. The explant was
pulsed-labeled with 3 μCi of [³H]proline 24 hours before harvesting. A high density of silver grains
appears over the smooth muscle cells in the medial wall. There are few silver grains over the cells
within the intimal thickening which fills the upper two-thirds of the micrograph.

vessel–normal diet group, and 22.2 mU in the injured vessel–cholesterol diet group (see Table I). In the group with the cholesterol diet and vessel injury the fourfold increase in concentration of cathepsin D is in agreement with the results reported by Peters *et al.* (1972). The concentrations of cathepsin D after 6 days in culture were found to be 4.3 mU/mg wet weight in the control–normal diet group, 5.0 mU/mg wet weight in the injured–normal diet group, and 21.8 mU/mg wet weight in the cholesterol diet–injured group. The culturing of intact explants of normal and injured vessel did not alter greatly the concentration of cathepsin D in the vessel. Hence, like the histological evidence, biochemical evidence suggests very little alteration with culturing.

2. COLLAGEN SYNTHESIS

Synthesis of new collagen by aorta explants was observed. The collagen fraction from labeled aorta explant homogenates was the material electrophoresed on a SDS-PAGE 6% slab gel. The stained gel showed bands in the $\alpha 1$ and $\alpha 2$ regions, which indicated the presence of type I collagen in the aorta. The fluorogram showed that radioactivity was present in these $\alpha 1$ and $\alpha 2$ bands (figure not shown). These results showed that the rabbit vessel in organ culture synthesized mostly type I collagen. However, the density of the $\alpha 1$ band in comparison to that of the $\alpha 2$ band indicated that type III collagen was present also. The presence of AB collagen, a basement membrane-like collagen found in whole aorta, was not detected as radioactive material in these organ culture preparations (Ehrlich and Trelstad, 1977), however, protein stained gels showed it to be present. We do not know why AB collagen is not actively synthesized in these organ cultures.

TABLE I

CATHEPSIN D ACTIVITY IN RABBIT AORTA HOMOGENATES

	Milliunits of Cathepsin D per milligram of tissue wet weight[a]	
Group	0 day in culture	6 days in culture
I: Control–normal diet	5.8	4.3
II: Injured vessel–normal diet	5.1	5.0
III: Injured vessel–cholesterol diet	22.2	21.8

[a] A unit of cathepsin D activity is a 1.0 change in optical density at $E = 280$ nm of a trichloacetic acid-soluble digest homogenate of aorta incubated at 40°C for 10 minutes. These are the average values from five separate rabbit experiments.

C.   Localization of Cathepsin D

The extracellular location of cathepsin D was searched for, using immunological antibody-trapping techniques. In the uninjured vessel–normal diet group,
there was no fluorescence to show extracellular secretion of cathepsin D in the
intima or medial wall. However, the adventitia showed the presence of extracellular cathepsin D (Fig. 4A). In the injured group receiving a normal diet, extracellular cathepsin D was not found in the intima, which was quite fibrous, or
in the medial wall but, as in the control group, extracellular cathepsin D was
found in the adventitia. In the injured vessel–cholesterol diet group, there was

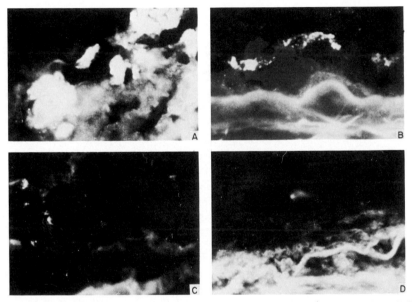

FIG. 4.   The extracellular trapping of cathespin D as an immune complex in rabbit aorta. (A) The
adventitia of a normal rabbit aorta, showing fluorescence over the fibroblasts. The aorta was incubated for 2 days with DMEM with normal rabbit serum supplemented with 5 mg/ml sheep anti-rabbit
cathepsin DF(ab)$_2$. This medium was replaced with serum-free DMEM, and the incubation continued
for an additional 24 hours. The explants were frozen, sectioned, fixed, stained for sheep IgG, and
photographed through a fluorescence microscope. (B) The intimal thickening of a scraped aorta from a
rabbit maintained on a 1% cholesterol diet for 3 months. The conditions for culturing and processing
the explant are identical to those described in (A). At the bottom of the figure are autofluorescent
elastic fibers of the medial wall. At the lumenal edge of the lesion is an extracellular cathepsin
D–antibody complex. In the center of the lesion no cathepsin D–antibody complex present. (C) The
same as (A), except that normal sheep IgG was substituted for sheep anti-rabbit cathepsin D IgG
during organ culturing. The only fluorescence observed is that from the autofluorescence of the
elastic fibers. (D) The same as (B), except that normal sheep IgG was substituted for sheep anti-rabbit
cathepsin D IgG during the culturing period. The autofluorescence of the medial wall elastic fibers is
seen.

fluorescence in both the adventitia and atheromatous-like plaque; the cathepsin D extracellular location was surrounding the cell closest to the lumen (Fig. 4B). These were the same cells from the autoradiographic studies that showed the highest synthetic activity. Cells in the deep lesion, those closest to the medial wall, showed no immunofluorescence activity, i.e., extracellular cathepsin D secretion. It appears that in the atherosclerotic lesion the high increase in cathepsin D might be related to new cells that are rapidly growing and synthesizing proteins at the edge of the growing atheroma.

D. Human Study

In a pilot study a piece of mesentary artery from a human surgical specimen was cultured for 9 days, [³H]proline was added on day 8, and autoradiograph studies showed good synthesis of protein in the medial wall and adventitia. In the thickened intimal plaque there was little synthetic activity (see Fig. 3D), indicating that the lesion was not as synthetically viable as the medial wall or adventitia. Some labeled vessels were homogenized and pepsinized. The salt-insoluble collagen fraction was run on SDS-PAGE gels, and fluorograms were made from these gels, as described in Section II. The presence of type I and type III collagen was found in this tissue, but no AB collagen was synthesized. Normal human vessels can be maintained in organ cultures for studying the growth of the extracellular matrix. In the future we wish to study the growth of atherosclerotic plaques in human arteries.

IV. Discussion

The organ-cultured artery is more complicated compared to a monolayer culture of endothelial cells or smooth muscle cells but is less complicated compared to the intact vessel *in vivo*. The whole vessel in culture has cells that have interactions with one another as well as interactions with connective tissue matrix. The organ culture environment can be as rigidly controlled as that of monolayer cultures. For the most part, tissues and organs in culture rely upon the diffusion of nutrients for survival. Since arteries are maintained for the most part by diffusion *in vivo*, they do quite well in organ culture. Cells in monolayers have a much greater capacity to survive in culture, as compared to whole organs, because the diffusion of nutrients involves only a single cell layer. In organ culture, an intact artery with its various cell populations will be useful for short-term studies requiring a controlled environment.

Monolayer cell cultures are useful for long-term studies with a single cell population in a controlled environment. In monolayer cell cultures cells that do not grow rapidly are selected out. Cells that grow out from explants or attach well to a petri dish surface will be the cells studied in a monolayer culture. In arterial organ cultures the cells which would be expected to grow out from such explants are prevented from doing so by using either Millipore filter surfaces or stainless steel mesh surfaces. The organ culture vessel works well because this organ retains its histological structure in the controlled environment. There is a possibility that the cells found to secrete cathepsin D and synthesize the most proteins in the intimal lesion may not have been the cells that grow out to form monolayer cell cultures. Hence the cells responsible for increases in cathepsin D synthesis and protein synthesis in intimal lesions may be lost in monolayer culture. The detection of their *in vivo*-like cellular function may be limited to organ cultures.

Stationary organ cultures were employed for the identification of cells that release cathepsin D into the extracellular matrix. With the use of this organ-culturing technique, a population of cells that secrete cathepsin D was identified in the atheromatous-like plaque and in the adventitia. In normal diet groups, injured and uninjured, there was no release of cathepsin D in the intima or medial wall, but there cathepsin D was released in the adventitia. In stationary cultures vessel homogenates from cholesterol diet–injured animals maintained the same cathepsin D concentration after 6 days in culture. The cells in organ cultures appear to be able to maintain the same concentration of cathepsin D in culture as they do *in vivo*.

The effect of serum from normal diet rabbits, and from rabbits fed a 1% cholesterol diet for 3 months, on the synthesis and secretion of cathepsin D was not significantly different histologically or biochemically. Likewise, the effects of these two different sera did not seem to affect the rate of protein synthesis, as judged by autoradiographs. Whether the use of sera at higher concentrations, say 20 or 30% (St. Clair and Harpold, 1975), affects the synthesis and secretion of cathepsin D or the synthesis of protein was not investigated at this time.

In monolayer tissue cultures an individual cell type such as smooth muscle cells or endothelial cells can be studied as a single cell population (Jaffee *et al.*, 1976; Rauterberg *et al.*, 1977). In an organ culture these cell–cell interactions or cell–extracellular matrix interactions may alter behavior. As an example, smooth muscle cells in tissue culture (Barnes *et al.*, 1976) appear to produce type I and type III collagen in a different proportion compared to that measured in normal or atherosclerotic vessels (McCullagh and Balian, 1975). In our experiments type I and type III collagen were synthesized *in vitro*. The absence of AB collagen synthesis in our preparation was puzzling, since it was readily isolated from the vessels' extracts. There may be some environmental influence on the type of collagen a cell will make in cell culture and in organ culture. If the environment involves cell–cell interactions or cell–matrix interactions *in vivo,* the organ cultures may be a more productive way of studying changes in collagen synthesis in disease. Since adult aorta can be organ-cultured for more than a week, this may be a better way of studying the changes in the synthesis of different collagen types in injured and uninjured vessels, or in the developing or regressing atherosclerotic lesions.

The growth or regression of atherosclerotic lesions in organ cultures would be a way of studying this pathological condition. The effects of chemicals or cell products could be evaluated using this culturing technique. Organ cultures of scraped whole aorta may be the way to investigate cellular proliferation of smooth muscle cells from the medial wall into the intimal space in a controlled environment. The use of vessels with developed atherosclerotic-like lesions may be useful for investigating regression, necrosis, or calcification of intimal plaque in a rigidly controlled environment.

Each of the two organ culture systems used to study aortas *in vitro* has its own advantages. In stationary cultures simple equipment can be used, and a great number of artery segments can be studied at a single time. Their disadvantage is that there is very little tissue in each explant for biochemical studies. The roller culture technique requires a great deal more time and equipment to set up, but there is more tissue per culture and optimal mixing of the medium with explants. In both culturing systems the aortas can be kept in good, metabolically active states and histologically healthy conditions for 10–12 days in a controlled environment. Longer periods of time may be possible with the use of another medium such as Leibovitz L-15 (see Blacklow *et al.*, 1975). The use of embryonic arteries may also increase the longevity of the organ cultures.

V.   Perspectives

At the current time roller cultures are being used to investigate the collagen types synthesized in rabbit aorta 1 week after balloon injury. By comparing the collagen types synthesized in newborn rabbit aorta versus normal adult rabbit aorta versus balloon-injured adult rabbit aorta, a pattern may develop that shows the relationship between the collagen types synthesized in vessel growth and the collagen types synthesized in intimal injury.

Roller cultures may be useful for the study of regression and necrosis in human atherosclerotic vessels. Organ culture is a way that synthetic activity of intact vessels can be followed with human arteries in a controlled environment. Vessels can be obtained from tissue removed at surgery or from fresh autopsy material. These vessels can be maintained in roller culture, and biochemical changes in normal or diseased vessels in a controlled environment can be compared. The deposition of calcium may also be studied using organ cultures. The incorporation of calcium-45 into normal and intimal thickened vessel can be studied.

The use of artery explants as a way of investigating atherosclerosis and other vascular diseases has not often been used. Arteries in organ culture may be a very useful system for studying human disease. The idea that arteries live *in vivo* by diffusion and lack an elaborate circulatory system for their nutrition and maintenance makes them very good tissue for organ culture. The maintenance of vessels *in vitro* for a 10-day period may be enough to identify alterations in metabolism or morphological structure in blood vessels.

ACKNOWLEDGMENTS

I wish to thank Drs. David Morton, Alan Barrett, John Dingle, Werner Jacobson, and Robin Poole and Ms. Ros Hembry of Strangeways Research Laboratory for their help and advice and the reagents involved in this work. I would also like to thank Deborah Scharf, Barbara Feldman, and Kimiko Hayashi for their help in preparing the manuscript.

REFERENCES

Barnes, M. J., Morton, L. F., and Levine, C. I. (1976). *Biochem. Biophys. Res. Commun.* **70,** 339–347.

Barrett, A. J. (1972). *In* "Lysosomes: A Laboratory Handbook" (J. T. Dingle, ed.), p. 123. North-Holland, Publ., Amsterdam.

Blacklow, N. R., Rose, F. B., and Whalen, R. A. (1975). *J. Infect. Dis.* **131,** 575–578.

Bonner, W. M., and Laskey, R. A. (1974). *Eur. J. Biochem.* **46,** 83–88.

Ehrlich, H. P., and Trelstad, R. L. (1977). *Circulation* **56,** 332.

Fell, H. B., and Robison, R. (1929). *Biochem. J.* **23,** 767–784.

Fritz, K. E., Augustyn, J. M., Peters, T., Jarmolych, J., and Daoud, A. S. (1976). *Atherosclerosis* **23,** 177–190.

Jaffee, E. A., Minick, C. R., Adelman, B., Becker, C. G., and Nachman, R. (1976). *J. Exp. Med.* **144,** 209–225.

Lowry, O. H., Rosebrough, N. J., Farr, A. L., and Randall, R. J. (1951). *J. Biol. Chem.* **193,** 265–274.

McCullagh, K. A., and Balian, G. (1975). *Nature (London)* **258,** 73–75.

Peters, T. J., Muller, M., and deDuve, C. (1972). *J. Exp. Med.* **136,** 1117–1139.

Poole, A. R., Hembry, R. M., and Dingle, J. T. (1974). *J. Cell Sci.* **14,** 139–161.

Poole, A. R., Hembry, R. M., Dingle, J. T., Pinder, I., Ring, E. F. J., and Cash, J. (1976). *Arthritis Rheum.* **19,** 1295–1307.

Rauterberg, T., Allan, S., Brehmer, U., Wirth, W., and Hauss, W. H. (1977). *Hoppe-Seyler's Z. Physiol. Chem.* **358,** 401–407.

Sade, R. M., Folkman, J., and Cotran, R. A. (1972). *Exp. Cell Res.* **74,** 299–306.

St. Clair, R. W., and Harpold, G. T. (1975). *Exp. Mol. Pathol.* **22,** 207–219.

Stemerman, H. B., and Ross, R. (1972). *J. Exp. Med.* **136,** 769–789.

Strangeways, T. S. P., and Fell, H. B. (1926). *Proc. R. Soc. London, Ser. B* **99,** 340–366.

Trowell, O. A. (1956). *Exp. Cell Res.* **16,** 118–147.

Weston, P. D. (1969). *Immunology* **17,** 421–428.

Chapter 9

Human Endothelial Cells in Vitro[1]

GARY E. STRIKER, JOHN M. HARLAN,
AND STEPHEN M. SCHWARTZ

Departments of Pathology and Medicine,
University of Washington,
Seattle, Washington

[1]Supported in part by NIH grants GM-21797, HL-11775, HL-03174 and HL-18645.

I.   Introduction

Arterial vascular disease accounts for substantial morbidity and mortality in humans. In many of these conditions endothelial injury has been postulated to be an important component. Thus, there have been numerous attempts to propagate endothelial cells *in vitro* to study suspected pathogenic factors in a systematic fashion under conditions where individual variables can be manipulated. Endothelial cells are metabolically quite active and synthesize or modify a number of biologically important substances. While considerable success has been reported for bovine (Gospodarowicz *et al.*, 1978), porcine (Slater and Sloan, 1975), ungulate (Blose and Chacko, 1975), and rodent arterial endothelial cells, the isolation and propagation of human arterial endothelial cells has been difficult. In fact, only endothelial cells from umbilical vein have been regularly obtained as cell strains that can be passaged. These cells have proven to be useful, but interpretation of data generated with them has several limitations; they are fetal cells, are venous in origin, are obtained from a unique blood vessel, and may be preprogrammed for senescence. These facts, coupled with the presumed importance of the arterial endothelium in the genesis of acute and chronic vascular disease, make it imperative that continuous vigorous attempts be made to identify the *in vitro* conditions necessary for the successful cultivation of human arterial endothelial cells.

II.   Identification and Characteristics of
Endothelial Cells *in Vitro*

A.   General

The identification of cells *in vitro* is often difficult, depending on the maintenance of *in vivo* functional and/or morphological traits. Initially, endothelial cells were thought to have few such markers. However, further investigation has shown that they possess several unique morphological, functional, and immunological features which are useful as markers.

B.   Morphology

Dilute plated endothelial cells initially have an elongated, angular shape. As they become confluent, they become more polygonal in shape and at confluence appear as a regular monolayer of cells that do not multilayer (Fig. 1).
Electron microscopic examination of human umbilical vein endothelium reveals these cells to contain Weibel–Palade (1964) bodies characteristic cytoplas-

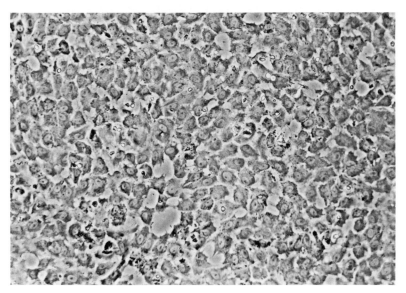

FIG. 1.   At confluence umbilical vein endothelial cells have a uniform polygonal shape.

mic organelles. By immunofluorescence and electron microscopy, endothelial cells contain only a small number of 60-Å thin filaments (actin), but 100-Å intermediate filaments are prominent, particularly around the nucleus (Haudenschild *et al.*, 1975).

C.   Thromboresistance

A feature characteristic of endothelial cells is that they present a nonthrombogenic surface to platelets *in vitro* (Harker *et al.*, 1977). When a suspension of endothelial cells is incubated with platelet-rich plasma at 37°C, no aggregation of platelets occurs (Fig. 2). These data contrast sharply with those reported for smooth muscle cells and fibroblasts, which cause prompt aggregation. This property was not a function of cell viability, since freeze-thawed preparations of both endothelial and control cells gave the same results as intact cells. The thromboresistance of endothelial cells was not altered by prostaglandin inhibitors, suggesting that this phenomenon was not dependent on prostacyclin (PGI_2), but probably was an intrinsic property of the endothelial cells.

Endothelial cells also contribute to the nonthrombogenicity of the vessel wall by the synthesis of PGI_2 (Weksler *et al.*, 1977; Moncada *et al.*, 1977). It was found that, when endothelial cells were admixed with smooth muscle cells, the aggregation of platelets *in vitro* was inhibited (Harker *et al.*, 1977). The supernate of endothelial cell cultures had a similar effect. In this case the phenomenon

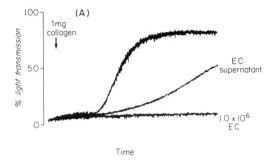

FIG. 2A.   Inhibition of colalgen-induced paltelet aggregation by PGI$_2$ synthesized in endothelial cells. Baseline platelet aggregation and release result from the addition of 1 mg of collagen preparation to platelet-rich plasma in an aggregometer system as measured by increasing light transmission (upper curve). When cultured human endothelial cells (EC) are included in the platelet suspension at a concentration of 1×10^6 cells/ml, collgen-induced platelet aggregation and release are completely inhibited (lower curve). The supernatent endothelial cell culture medium also inhibits collagen-induced platelet aggregation (middle curve).

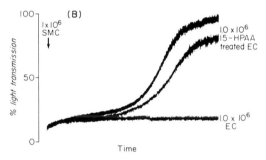

FIG 2B.   Arterial smooth muscle cell (SMC)-induced platelet aggregation and release. Cultured human medial smooth muscle cells from human arteries initiate platelet aggregation and release when added to a spun platelet-rich plasma system, as shown by increasing light transmission (upper curve). Inhibition is complete when cultured endothelial cells are included in the suspension (lower curve). Preincubation of endothelial cells with 15-hydroperoxyarachidonic acid (15-HPAA) abolishes the capacity of endothelial cells to inhibit platelet aggregation and release (middle curve).

was inhibited by indomethacin and required the addition of either fresh supernate from endothelial cell cultures or viable endothelial cells to a mixture of smooth muscle cells to inhibit platelet aggregation. Weksler *et al.* (1977) have described a method for quantitating PGI$_2$ release from human endothelial cells *in vitro* using the inhibition of thrombin-induced release of aspirin-treated platelets. PGI$_2$ production, although characteristic of endothelial cells, has also been described in other cell types, albeit at lower levels.

In summary, endothelial cell thromboresistance appears to depend both on intrinsic properties of the endothelial cell surface and of a short-lived humoral antithrombogenic agent, PGI_2. The relative contributions of these two mechanisms *in vivo* remain unknown.

D.   Factor VIII

Factor VIII is synthesized by endothelial cells (Jaffe *et al.*, 1973b) and is one of the most useful markers for endothelial cells *in vitro*. Morphological studies of the body distribution of factor VIII reveal that its main localization is in the region of the vascular intima (Hoyer *et al.*, 1973). Corroborating these data is the *in vitro* finding that human umbilical vein cells contain this antigen, whereas overgrowing smooth muscle cells, or several other cell types tested, do not. The exact localization of factor VIII is not known, although its immunofluorescence appearance suggests that a significant portion is contained within cytoplasmic vesicles. Thus factor VIII antigen appears to be a very useful tool for differentiating endothelial cells from other cells that might potentially contaminate cultures derived from vascular intima.

One note of caution is that, in human pulmonary artery (Johnson and Erdos, 1978) and bovine aortic endothelial cell cultures, an elongated "fibroblastoid" or "sprout" cell type appears late in culture. These cells demonstrate factor VIII antigen localization. In the case of the sprout cells from bovine aortic endothelial cell cultures, they have been shown to synthesize collagen types I, III, A, and B in medium and/or cell layers. Before sprouting occurred, no type I collagen was found (Cotta-Pereira *et al.*, 1979). Bovine smooth muscle cells, in parallel experiments, synthesized collagen types which were indistinguishable from sprout cells. Thus cells resembling smooth muscle cells may appear, late in culture, even after many passages. The confounding fact is that, while these cells still possess the factor VIII antigen marker, they have altered morphology and protein synthetic patterns resembling those of vascular smooth muscle cells.

E.   Angiotensin-Converting Enzyme

An additional feature characteristic of endothelial cells is the presence of the cell surface enzyme angiotensin-converting enzyme (Smith and Ryan, 1972). This enzyme is responsible for the conversion of angiotensin I to angiotensin II and the activation of bradykinin. This reaction principally occurs in the pulmonary circulation, but the enzyme has been found on all endothelial surfaces examined and in renal cortex. The enzyme appears to be a useful marker for identifying endothelial cells (Caldwell *et al.*, 1976).

F.   Fibronectin

Fibronectin or cold insoluble globulin, a 240,000 molecular-weight plasma protein, is synthesized by many cells in the body. Endothelial cells have recently been postulated to be the major source of the fibronectin found in plasma (Jaffe and Mosher, 1978). In view of the role postulated for fibronectin as an opsonin and as a cell attachment factor, this property of endothelial cells may prove to be of importance in the pathophysiology of sepsis and wound healing.

G.   Collagen Synthesis

The synthesis of collagenous proteins by endothelial cells has been reported (Howard *et al.*, 1976; Jaffe *et al.*, 1976). There is an apparent difference between that synthesized by bovine aortic and human fetal umbilical cells. The former have been shown to synthesize collagen type III and an unidentified collagen with a molecular weight of 115,000–120,000 (Cotta-Pereira *et al.*, 1979). In contrast, bovine smooth muscle cells synthesize collagen types IV, III, I, A, and B under similar conditions. Human endothelial cells resembled bovine endothelial cells, but in addition synthesized type IV collagen (Jaffe *et al.*, 1976; Howard *et al.*, 1976; P. K. Killen and G. E. Striker, unpublished observations, 1979) and a large amount of $[\alpha 1(I)]_3$ (P. K. Killen and G. E. Striker, unpublished observations, 1979). Both endothelial cell strains can be distinguished from smooth muscle cells by the appearance of $\alpha 2$ collagen chains in pepsin-digested medium or cell layers. This serves as a marker for the presence of contaminating smooth muscle cells in an endothelial cell culture (P. K. Killen and G. E. Striker, unpublished observations, 1979) (Fig. 3).

H.   Surface Markers and Receptors

The presence of ABH antigens on cultured human endothelial cells was first reported by Jaffe *et al.* (1973a). Moraes and Stasny (1977) noted HLA-ABC and Dr antigens and non-HLA antigens shared with blood monocytes (E), but not lymphocytes. C3b receptors were not found on human umbilical vein endothelial cells by Scheinman *et al.* (1978), and in our laboratories neither C3b nor Fc receptors were demonstrable under controlled conditions (Killen *et al.*, 1978). Binding of low-density lipoprotein in bovine aorta (Vlodavsky *et al.*, 1978), thrombin in human umbilical vein (Awbrey *et al.*, 1975), heparin in human umbilical vein (Glimelius *et al.*, 1978), estrogen in rabbit aorta (Colburn and Buonassisi, 1978), and factor VIII in human (R. T. Wall, R. Count, L. A. Harker, and G. E. Striker, unpublished observations, 1979) endothelium have been reported.

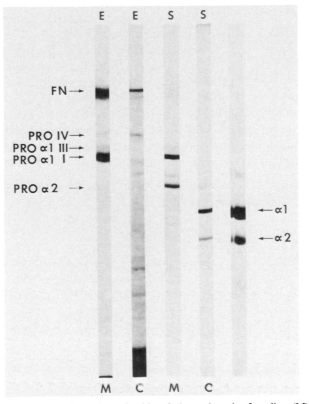

FIG. 3.   Sodium dodecyl sulfate polyacrylamide gel electrophoresis of medium (M) and cell layer (C) proteins from cultures of human umbilical vein endothelial cells (E) and vascular smooth muscle cells (S). Endothelial cells differ from smooth muscle cells in that there are no α2 collagen chains present in either the medium or the cell layer.

III.   Long-Term Culture

A.   General

Since the description of successful culture of human umbilical vein endothelial cells by Lazzarini-Robertson (1959), Maruyama (1963), Fryer *et al.* (1966), Jaffe *et al.* (1973a,b), McDonald *et al.* (1973), Gimbrone *et al.* (1974), and Lewis *et al.* (1973), numerous laboratories have cultured human umbilical vein endothelial cells. Several problems have emerged:

1. It has been difficult to maintain these human umbilical vein cells in long-term culture with a high proliferative capacity. While an occasional isolate can be

passaged up to 10–12 times, the vast majority of isolates appear to senesce rapidly and seldom maintain a reasonable growth fraction after 2 or 3 passages.

2. Smooth muscle cell contamination often occurs with repeated passage of endothelial cells. Attempts to use plasma-derived serum deficient in platelet mitogenic activity or "thymidine suicide" to eliminate smooth muscle cell contamination have met with some success in our laboratories (Schwartz, 1978).

3. Human adult endothelial cells from other sources have not been successfully maintained for any length of time in culture, save for those described in a recent report on pulmonary endothelium (Johnson and Erdos, 1978). Similarly, primate endothelium has been exceedingly difficult to maintain and passage.

These problems all reflect the necessity of defining the appropriate growth conditions for human endothelial cells.

B.   Plasma and Serum Requirements

Human endothelial cells, like all mammalian cells, require plasma or serum for continued proliferation. In serum-free media at sparse density the cells become quiescent and will not proliferate. Increasing serum concentrations up to 50% enhances proliferation; both homologous and heterologous sera have been used successfully. In contrast to smooth muscle cells, which require serum with platelet-derived growth factors (Ross *et al.*, 1974), both R. T. Wall, R. Counts, L. A. Harker, and G. E. Striker, (unpublished observations, 1979) and Thorgeirrson and Robertson (1978) have demonstrated that umbilical vein endothelial cells grow well in plasma-derived serum which is deficient in platelet mitogenic activity.

C.   Growth Control

Endothelial cells have a characteristic cell density (Schwartz *et al.*, 1979). Cultures at this density will not respond to fresh serum or growth factors with initiation of DNA synthesis. This fact requires that growth studies be performed on subconfluent cultures measuring variables such as time to confluence, cell number at a given time in culture, and the ability to proliferate in the presence of limiting serum or other culture conditions.

Most cell types require one or more growth factors, some of which have been defined as polypeptides. Those identified include fibroblast growth factor (FGF), epithelial growth factor (EGF), platelet-derived growth factor (PDGF), insulin, and somatomedins. Endothelial cell response to several growth factors has been studied *in vitro,* and the findings have been somewhat variable. In our laboratories, FGF had only a modest growth-promoting effect on human umbilical vein endothelial cells (Wall *et al.,* unpublished observations, 1979). PDGF was also found to have little effect on human endothelial cells (Wall *et al.*, unpublished observations, 1979; Thorgeirrson and Robertson, 1978), although there are reports to the contrary (Fig. 4).

Gospodarowicz *et al.* (1978) found a potent mitogenic response to added FGF and thrombin utilizing a bovine aortic endothelial cell line that had been cloned in the presence of FGF. The effect on bovine aortic endothelial cells not previously exposed to FGF was minimal in our laboratories (Schwartz *et al.*, 1979). The latter cells were also independent of EGF, insulin, and PDGF.

In human umbilical vein endothelial cells, Gospodarowicz *et al.* (1978) noted a dramatic mitogenic response only when FGF or EGF was used simultaneously with thrombin. Under these conditions proliferation occurred even at a low plating density. In contrast, Maciag *et al.* (1979) failed to find mitogenic activity using purified FGF and endothelial cells that had never previously been exposed to FGF.

The reasons for the variability between laboratories are unclear. The report by Maciag *et al.*, (1979) may shed some light on the FGF controversy. These workers were able to find no mitogenic activity using high concentrations of purified FGF and human thrombin. They instead have described an extract of bovine hypothalamus that is a heat-labile, protease-sensitive, nondialyzable protein of 75,000 molecular weight, which has significant mitogenic activity for human umbilical vein endothelial cells. This endothelial cell growth factor was separable from fibroblast growth factor-like mitogenic activity by gel exclusion chromatography. Proteases such as thrombin were not required for its effect. The precise role of these pituitary and hypothalamic growth factors in controlling endothelial cell proliferation clearly requires further definition.

D.   Endothelial Cell Mitogens

Conditioned media have often been used to enhance cell growth *in vitro* and can frequently initiate DNA synthesis and one or two cell divisions. Wall *et al.* (unpublished observations, 1979) noted a slight enhancement of human endothelial cell proliferation with macrophage-conditioned media, and similar results have been reported using bovine arterial endothelium and wound fluid (Greenburg and Hunt, 1978). Gospodarowicz *et al.* (1978) have reported significant growth-promoting activity with media conditioned by virally transformed 3T3 cells. Recently, Fass *et al.* (1978) and Gajdusek *et al.* (1980) described 3T3 mitogenic activity using extracts from porcine endothelial or bovine endothelial cell-conditioned media (Fig. 5). The cocultivation of bovine endothelial cells and vascular smooth muscle cells in serum lacking PDGF resulted in an increased rate of replication of the smooth muscle cells. In the absence of endothelial cells, no proliferation was seen under these conditions, suggesting that the endothelial

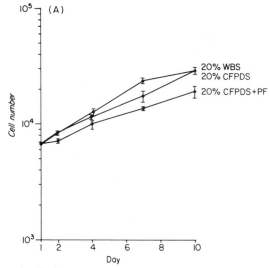

FIG. 4A.   Effect of CFPDS and WBS on time to confluent endothelial cell density. Umbilical vein endothelial cells plated initially at a concentration of 10,000 cells/well in 10% CFPDS and changed to 20% CFPDS, WBS, or CFPDS + PF after 24 hours. The medium was changed every 3 days. Each point represents the mean count of cells from four replicate wells ± one standard error.

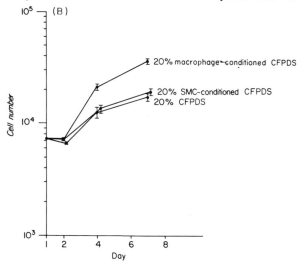

FIG. 4B.   Effect of macrophage-conditioned medium and smooth muscle cell (SMC)-conditioned medium on endothelial cell proliferation. Umbilical vein endothelial cells plated initially at a concentration of 10,000 cells/well in 10% CFPDS and changed to either 20% CFPDS or 20% CFPDS conditioned by human peritoneal macrophages or SMC (1×10^6 cells/10 ml medium). The medium was changed every 3 days. Each point represents the mean count of cells from four replicate wells ± one standard error. Macrophage-conditioned medium significantly augmented endothelial cells compared with fresh, untreated CFPDS and SMC-conditioned CFPDS at days 4 and 7 ($p < 0.01$).

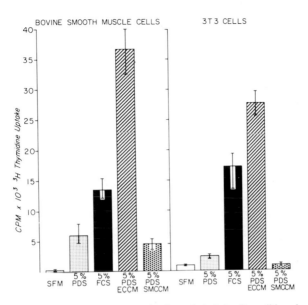

FIG. 5.   Stimulation of thymidine incorporation by endothelial cell conditioned media. 3T3 cells were plated at 5×10^3 cells/cm^2 in Falcon multiwells and brought to quiescence. Medium was replaced with 200 μl test media diluted with 300 μl fresh serum free medium. Thymidine incorporation was measured at 18 to 20 hours with 1uCi/ml [^{3}H]thymidine. SFM: serum free medium; 5% PDS: medium containing 5% PDS; 5% FCS: medium containing 5% fetal calf serum; 5% PDS ECCM: 5% PDS containing medium which was preincubated with endothelial cells for 48 hours; 5% PDS SMCCM: 5% PDS which was preincubated with smooth muscle cells for 48 hours.

cells were responsible for the mitogenic activity (Gajdusek *et al.*, 1980). The mitogenic activity of endothelial cell medium (serum-free) is quite high, being equipotent to 15% serum. Additionally, bovine vascular smooth muscle cells can be grown in PDGF-free serum if the serum is first incubated with bovine endothelial cells. Growth is equal to or exceeds that in equivalent concentrations of serum containing PDGF.

A great deal remains to be learned about the identity and effects of this factor(s) and its distribution among species as well as in various vascular beds within species.

IV.   Isolation of Endothelial Cells

The conditions for the initial isolation of umbilical vein endothelial cells were first described by Maruyama (1963) and then by Fryer *et al.* (1966). The ability to identify and passage human endothelial cells was reported by Jaffe *et al.* (1973b) and Gimbrone *et al.* (1974). Umbilical cords are collected within 12 hours of delivery, and the vein is cleared of blood by perfusing with medium 199. The most useful enzyme has proven to be a crude collagenase preparation (Calbiochem, B grade, 1 mg/ml) in phosphate-buffered saline, pH 7.4. Purified collagenase preparations give a poor yield, and trypsin apparently injures the cells more than crude collagenase preparations. These observations suggest that enzymes in the crude preparation other than collagenase are required for the removal of endothelial cells from the vascular wall. The cord containing the enzyme is incubated at 37°C for 30 minutes, and the vessel is then perfused with several volumes of Hanks' balanced salt solution. Some authors have stressed careful handling of the cord to avoid trauma, whereas others deliberately compress or massage it to remove endothelial cells. A pellet is formed from the perfusate by centrifuging at 200 *g* for 5 minutes. The cells are suspended in 20% serum in complete Waymouth's medium and transferred to flasks in a gassed incubator. The amount and source of serum has varied from laboratory to laboratory. Initially, all investigators utilized fetal calf serum. Subsequently, it was stated that pooled human serum was a better serum source (Gospodarowicz *et al.*, 1978). We have employed both and find that, if serum lots are screened, fetal calf serum is as effective a serum source as pooled human serum.

The cells are passaged by incubating the cell layer at 37°C with a 0.05% trypsin solution in EDTA for 2 minutes, or until the cells are detached. The trypsin is inactivated by the addition of Waymouth's medium containing 20% serum. Although it appears that the cells must be plated at high density to ensure survival and proliferation, this point has not been systematically studied.

Nonprimate Endothelial Cells

Many laboratories have reported successful culture of endothelial cells from nonprimate species using both fetal and adult sources and both venous and arterial sites. Their usefulness, however, may be limited in the study of human diseases. The difficulty in growing human endothelial cells from both umbilical vein and other sites suggests that its growth requirements may be considerably different. For example, Gospodarowicz *et al.* (1978) reported that human umbilical vein endothelial cells contained a receptor for EGF, while bovine aortic endothelial cells did not. Similar differences in response to growth factors or conditions would be expected. In addition, the use of certain nonhuman species for the study of human diseases such as atherosclerosis must always be suspect, especially when the species does not develop spontaneous atherosclerosis. Even the use of fetal human venous endothelium for studies on diseases such as atherosclerosis that are limited to arterial sites is worrisome.

The potential importance of the problem of species and site of origin is illustrated by our studies with bacterial endotoxin [J. H. Harlan, L. A. Harker, and G. E. Striker (unpublished observations, 1979)]. Because of the potential importance of endothelial injury in the pathogenesis of septic shock and its sequelae, we examined the effect of bacterial lipopolysaccharide on vascular endothelium using a chromium release assay for lethal lytic injury and a detachment assay for sublethal injury. Human endothelial cells were not affected in either assay with bacterial lipopolysaccharide concentrations up to 100 μg/ml. In contrast, a sublethal, nonlytic injury to bovine aortic endothelial cells is detectable at bacterial lipopolysaccharide concentrations as low as 10 pg/ml. The striking differences in response to bacterial lipopolysaccharide may reflect species difference, adult versus fetal origin, or arterial versus venous sites. Similar questions may pertain to any other processes studied unless endothelial cells are available from the appropriate site and species.

V.   Perspectives

A.   General

The only endothelial cell from humans that can be regularly cultured is from umbilical vein endothelium. There is, however, fairly extensive experience with aortic endothelium from four nonprimate species. Cells from these species can be readily cultured and have, at least in the case of bovine endothelium, a finite life span of as many as 30 generations (Schwartz, 1978) It seems reasonable that the difficulties in passaging adult primate arterial endothelium are technical rather than fundamental to the biology of these cells.

B.   Cooperativity

A number of conditions may be important for passaging human endothelial cells. First, there is the possibility that endothelial cell growth requires a self-mitogen (Gajdusek *et al.*, 1980). Endothelial cell products such as proteases and prostaglandins have been shown in other cell systems to affect proliferation. The role of these products needs to be defined in the initiation and modulation of human endothelial cells *in vitro*. For instance, they may be important in the phenomenon of cooperativity (Atherton, 1977). As shown by Dulbecco and Elkington (1973) for epithelial lines with growth properties in many ways similar to those of endothelium, the growth of certain cells in culture may depend on the availability of near neighbors (see also Atherton, 1977).

If endothelial cells produce a mitogen, consideration of the relationships of endothelial cells and smooth muscle cells *in vivo* becomes considerably more complex. This poses the question of whether mitogen production might be modulated at sites of vascular injury. A second question is raised by the properties of endothelial cell-conditioned medium (Gajdusek *et al.*, 1980). Endothelial cell-conditioned medium added to 0.2% serum supported the growth of endothelial cells, although the same cells did not grow in ordinary medium supplemented with this concentration of serum. This is similar to the observations on tumor cell lines capable of growth under conditions of extremely low serum concentration, apparently because they produce self-growth factors (DeLarco and Todaro, 1975). It is also possible that the apparent independence of endothelial cells of the growth factors required by other cells is a result of the production of a self-mitogen. Currently there are no data on the identity of this factor, its resemblance to other defined growth factors, or its role in the intact animal.

In summary, there is evidence that endothelial cells produce a factor, or factors, that are mitogenic for other cells. The putative factor may also be a self-mitogen and be important in the isolation and propagation of endothelial cells.

C.   Substrate

It may also be important to consider the use of different kinds of substrates. This includes the chemical composition of natural substrates such as collagen and fibronectin and the nature of artificial substrates such as DEAE-Sephadex beads (Levine *et al.*, 1977). The bead shape, by allowing cells to grow onto a very large surface area, may ameliorate problems associated with passaging and culture at low density.

D.   Mode of Growth

Another important consideration is the mode of endothelial growth. Most cell cultures utilize cells that have been dispersed and are required to regrow to a confluent layer. *In vivo,* at least in the adult animal, this never happens with the endothelium. After denudation endothelial growth occurs only by regeneration from the edge of the intact cell sheet. In other situations, neovascularization occurs by outgrowths from existing vessels. Thus, the use of DEAE-Sephadex beads or other modifications of substrate may be useful in establishing human endothelial cells in culture and increasing the yield of cells by providing an appropriate surface geometry that more closely stimulates that *in vivo.*

E.   Growth Requirements

One possibility is that endothelial cell growth from different species is dependent on the specific composition of the medium, including such variables as oxygen tension, reduction potential of the medium, composition of low-molecular-weight nutrients, fatty acids, and growth factors. This poses the dilemma of a system with too many variables to test.

F.   Source of Endothelial Cells

Just as the umbilical vein endothelium appears to be relatively easy to establish in cell culture, it is conceivable that there are other vascular beds in the adult animal that would be equally easy to establish. There has been no careful study on the culture properties of other human vascular beds either in the newborn or in the adult. There are, however, intriguing preliminary reports which suggest that human pulmonary endothelium may be possible to culture (Johnson and Erdos, 1978).

G.   Contaminating Cell Types

There are two major problems in establishing endothelial cell cultures free of contamination by other cell types. First, there is the fact that the limitation of endothelial cell growth to the monolayer configuration means that other cell types are likely to proliferate and overgrow the culture. Second, the observation that endothelial cells themselves produce a mitogen that is active for other cells implies that these other cell types will have a growth advantage, particularly when endothelial cells are studied at confluent densities. It seems clear that techniques for purifying cultures will continue to be important, particularly as cells are isolated from more complex vascular beds.

H.   Identification

In this chapter we have stressed techniques that label the endothelial cell, particularly those detecting the presence of factor VIII antigen and angiotensin-converting enzyme antigen. These techniques, or any other techniques specific for the endothelial cell, are limited by the fact that they label the major cell type and not the contaminating cell type which by its presence in small numbers may represent a serious problem both for long-term culture and for the certainty that the results obtained in individual experiments represent the properties of the major type. It is thus extremely important to develop appropriate markers for the potential contaminating cell types. Two markers that have been suggested are the specific antibody for smooth muscle cell actin (Groschel-Stewart *et al.*, 1975) and the presence of specific collagen types (P. K. Killen and G. E. Striker, unpublished observations, 1979).

References

Atherton, A. (1977). *Cancer Res.* **37**, 3619-3622.

Awbrey, B. J., Owen, W. G., Fry, G. L., Cheng, F. H., and Hoak, J. C. (1975). *Blood* **46**, 1046.

Blose, S. H., and Chacko, S. (1975). *Dev., Growth Differ.* **17**, 153-165.

Caldwell, P. R. B., Seegal, B. C., Hsu, K. C., Das, M., and Soffer, R. L. (1976). *Science* **191**, 1050-1051.

Colburn, P., and Buonassisi, V. (1978). *Science* **201**, 817-819.

Cotta-Pereira, G., Sage, H., Schwartz, S. M., Bornstein, P., and Ross, R. (1979). *West Connect. Tissue Soc. Proc.*, p. 25a.

DeLarco, J. E., and Todaro, G. J. (1975). *Proc. Natl. Acad. Sci. U.S.A.* **75**, 4001-4005.

Dulbecco, R., and Elkington, J. (1973). *Nature (London)* **246**, 197-199.

Fass, D. N., Downing, M. R., Meyers, P., Bowie, E. J. W., and Witte, L. D. (1978). *Circulation, Suppl.* **11**.

Fryer, D. G., Birnbaum, G., and Luttrell, C. N. (1966). *J. Atheroscler. Res.* **6**, 151-163.

Gajdusek, C., DiCorleto, P., Ross, R., and Schwartz, S. M. (1980). *J. Cell. Biol.* **84**, (in press).

Gimbrone, M. A., Jr., Cotran, R. S., and Folkman, J. (1974). *J. Cell Biol.* **60**, 673-684.

Glimelius, B., Busch, C., and Hook, M. (1978). *Thromb. Res.* **12**, 773-782.

Gospodarowicz, D., Brown, K. D., Birdwell, C. R., and Zetter, B. R. (1978). *J. Cell Biol.* **77**, 774-788.

Greenburg, G. B., and Hunt, T. K. (1978). *J. Cell. Physiol.* **97**, 353-360.

Groschel-Stewart, U., Chamley, J. H., Campbell, G. R., and Burnstock, G. (1975). *Cell Tissue Res.* **165**, 13-22.

Harker, L. A., Striker, G. E., Wall, R. T., and Quadracci, L. J. (1977). *Clin. Res.* **25**, 515a.

Haudenschild, C. C., Cotran, R. S., Gimbrone, M. A., and Folkman, J. (1975). *J. Ultrastruct. Res.* **50**, 22-32.

Howard, B. V., Macarak, E. J., Gunson, D., and Kefalides, N. A. (1976). *Proc. Natl. Acad. Sci. U.S.A.* **73**, 2361-2364.

Hoyer, L. W., De los Santos, R. P., and Hoyer, J. R. (1973). *J. Clin. Invest.* **52**, 2737-2744.

Jaffe, E. A., and Mosher, D. F. (1978). *J. Exp. Med.* **148**, 1779-1790.

Jaffe, E. A., Nachman, R. L., Becker, C. G., and Minicki, C. R. (1973a). *J. Clin. Invest.* **52**, 2745-2756.

Jaffe, E. A., Hoyer, L. W., and Nachman, R. L. (1973b). *J. Clin. Invest.* **52**, 2757-2764.

Jaffe, E. A., Minick, C. R., Adelman, B., Becker, C. G., and Nachman, R. (1976). *J. Exp. Med.* **144**, 209-225.

Johnson, A. R., and Erdos, E. G. (1978). *Circulation, Suppl.* **11**, 108.

Killen, P. K., Striker, G. E., and Byers, P. H. (1978). *Fed. Proc., Fed. Am. Soc. Exp. Biol.* **37**, 642.

Lazzarini-Robertson, A. (1959). Thesis, Cornell Medical College, New York.

Levine, D. W., Wong, J. S., Wang, D. I. C., and Thilly, W. G. (1977). *Somatic Cell Genet.* **3**, 149-155.

Lewis, L. J., Hoak, J. C., and Fry, G. L. (1973). *Science* **181**, 453-454.

Maciag, T., Cerundolo, J., Ilsley, S., Kelley, P. R., and Forand, R. (1979). *Proc. Natl. Acad. Sci. U.S.A.* **76**, 5674-5678.

McDonald, R. I., Shepro, D., Rosenthal, M., and Booyse, F. M. (1973). *Ser. Haematol.* **6**, 469-478.

Maruyama, Y. (1963). *Z. Zellforsch. Mikresk. Anat.* **60**, 69-79.

Moncada, S., Higgs, E. A., and Vane, J. R. (1977). *Lancet* **1**, 18-21.

Moraes, J. R., and Stasny, P. (1977). *J. Clin. Invest.* **60**, 449-454.

Ross, R., Glomset, J., Kariya, B., and Harker, L. (1974). *Proc. Natl. Acad. Sci. U.S.A.* **71**, 1207-1210.

Scheinman, J. I., Fish, A. J., Kim, Y., and Michael, A. F. (1978). *Am. J. Pathol.* **92,** 147–154.
Schwartz, S. M. (1978). *In Vitro* **14,** 966–980.
Schwartz, S. M., Selden, S. C., and Bowman, P. (1979). *In* "Hormones and Cell Culture" (R. Ross and G. Sato, eds.), Vol. 6, pp. 593–610. Cold Spring Habor Conferences on Cell Proliferation, Cold Spring Habor, New York.
Slater, D. N., and Sloan, J. M. (1975). *Atherosclerosis* **21,** 259–272.
Smith, U., and Ryan, J. W. (1972). *Adv.Exp. Med. Biol.* **4,** 267–276.
Thorgeirrson, G., and Robertson, A. L., Jr. (1978). *Atherosclerosis* **30,** 67–78.
Vlodavsky, I., Fielding, P. E., Fielding, C. J., and Gospodarowicz, D. (1978). *Proc. Natl. Acad. Sci. U.S.A.* **75,** 356–360.
Weibel, E. R., and Palade, G. E. (1964). *J. Cell Biol.* **23,** 101–112.
Weksler, B. B., Marcus, A. J., and Jaffe, E. A. (1977). *Proc. Natl. Acad. Sci. U.S.A.* **74,** 3922–3926.

Chapter 10

Human Arterial Wall Cells and Tissues in Culture [1]

RICHARD M. KOCAN, NED S. MOSS, [2]
AND EARL P. BENDITT

Department of Pathology,
School of Medicine,
University of Washington,
Seattle, Washington

I.   Introduction

Human arterial wall explants and isolated smooth muscle cells (SMCs) have been examined by methods of cell culture for several purposes: (1) morphological phenotypic stability, (2) replicative characteristics and replicative life span,

[1] The original investigations reported here were supported by USPHS grant HD-03174.
[2] *Present address:* Department of Pathology, Marshall University, Huntington, West Virginia 25701.

153

(3) production of collagen of different types, (4) uptake and metabolism of lipo-proteins, (5) response to growth factors and to insulin, and (6) rates of mutation. The purpose of this presentation is to review briefly some of these observations and to direct attention to certian problems that need further exploration if we are to derive information relevant to human biology and human disease from studies of this kind of material.

Arterial wall SMCs derived from the media have been cultivated from tissues obtained at surgery or postmortem. In general tissues removed at surgery seem to be best suited for cultivation for several reasons: (1) They are from individuals not subject to terminal disease and events associated with dying; (2) prior to initiation of culture, they have not been subjected to long times postmortem; (3) the chances of infection of tissue during surgical procedures are substantially less than at autopsy. As we shall see, other factors may be of greater importance than the conditions under which tissue is removed. The age of the donor, the site from which tissue is taken, and the presence of optimal culture medium conditions all play a role in the success of tissue survival and of cell growth capacities.

II.　Methods of Cultivation

A.　Initiation of Cultures

Several different procedures have been used to obtain cells from vascular tissue for *in vitro* culture. Attached explants, free-floating fragments, and enzyme-dissociated tissue fragments have all been tried with varying degrees of success (see Appendix).

B.　Attached Explants

Tissue fragments 1–2 mm in diameter are generally used in this procedure. To obtain these fragments, tissue is aseptically removed from the desired organ and placed on a sterile petri plate. With the use of scalpel blades the tissue is minced into smaller and smaller pieces until the desired size (∼ 1 mm) is obtained. These pieces are then washed with buffered saline or balanced salt solution to remove any debris prior to initiating the cultures. The fragments can be added to culture flasks or dishes in a small volume of medium and allowed to settle for several days, at which time many will have attached to the surface of the culture vessel (Ross, 1971; Fischer-Dzoga *et al.*, 1973). For attachment and outgrowth to occur more quickly the pieces can be covered with a sterile coverslip, and cell growth will occur on both the culture dish surface and the coverslip in about 3–5 days. We frequently use the technique of gently removing all but a thin film of medium from the cultures after distributing the fragments evenly over the culture dish surface. Incubation under these conditions overnight in a humidified incubator followed by the addition of a normal volume of medium results in good attachment and cellular outgrowth in 3–5 days (Fig. 1).

It should be emphasized that when dealing with aortic tissue all the adventitia or connective tissue must be removed or excluded from the cultures. Since this tissue contains fibroblasts, which have a greater growth capacity than SMCs, they will overgrow the cultures if not excluded. This is easily done by pulling all the connective tissue off the outer artery wall with forceps prior to mincing the tissue.

C.   Free-Floating Explants

Jarmolych et al. (1968) and Rossi et al. (1973) have successfully used this procedure to study phenotypic modulation in vascular SMCs under various conditions of cultivation. The procedure requires 1- to 5-mm fragments of arterial wall tissue, which are placed in culture dishes with sufficient medium to prevent their settling to the bottom. These fragments are obtained exactly as described for attached explants. Cells will grow over the periphery of the tissue fragments and can be studied by sectioning the fragments or can be removed with 0.05% trypsin to produce large numbers of individual cells. After the cells have been removed, the tissue fragments can be placed back into growth medium and cells will again grow over their surface. This procedure can be repeated several times. When roller bottles are used to move the fragments and medium slowly, faster and more luxuriant growth of cells results.

D.   Enzymatic Dissociation (Primary Culture)

A procedure that has been tried with varying success for human fibroblasts and vascular SMCs is enzymatic dissociation of tissue. This method consists of finely mincing the tissue with a scalpel or scissors and washing away cell debris, blood, and damaged tissue with a buffered salt solution, followed by exposure to one or more enzymes capable of removing elements of the extracellular matrix. This is done by adding the minced tissue to the appropriate prewarmed enzyme and stirring the mixture with a magnetic stirrer for 10–15 minues. The cloudy, cell-rich supernate is then poured into a prechilled flask containing buffered balanced salt solution to stop the enzymatic reaction. The remaining tissue is again covered with warm enzyme and allowed to digest further. This procedure is repeated until essentially all the tissue is digested. EDTA (0.02%) plus trypsin (0.025%), 0.1% collagenase, 0.25% hyaluronidase, and 0.0125% elastase have all been successfully used on a variety of smooth muscle tissue including vascular tissue. After dissociation, the chilled cell suspension is centrifuged, the enzyme is removed, and then the cells are suspended in the appropriate medium at a

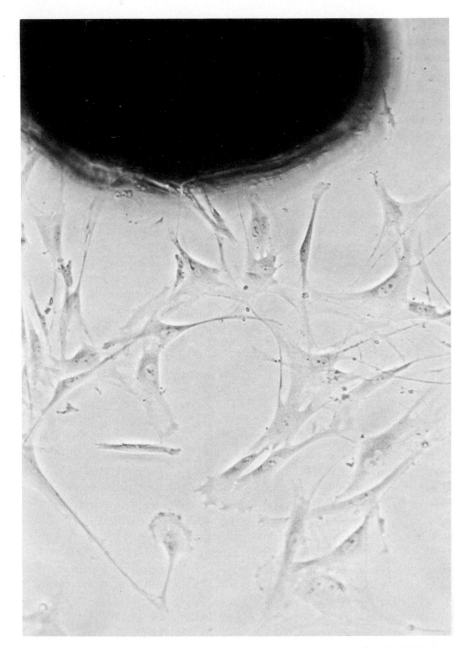

FIG. 1. An attached explant of human fetal aorta showing migration of the first few SMCs onto the culture dish. If refed with fresh medium over several weeks, the migrating cells will begin to replicate and eventually produce a monolayer of cells over the bottom of the dish with the aortic tissue fragments still attached. Culture medium: Eagle's MEM plus 10% FCS. Time: 5–10 days in culture. Magnification: ×100.

concentration suitable for culturing. This can be done by treating the cells with trypan blue and counting the number of viable cells on a hemocytometer. Living cells exclude the dye and appear bright, while dead cells take up the dye and appear dark. Cells in such a suspension take from several minutes to several hours to settle out, attach, and spread over the surface of the culture dish (Chamley *et al.*, 1977; Mark *et al.*, 1973; May *et al.*, 1975). In a comparison of the growth characteristics of SMCs derived from uterine walls, leiomyomas, femoral artery walls, and atherosclerotic plaques it was found that uterine but not arterial wall cells retained the capacity to proliferate after enzymatic dissociation. This was so despite the fact that the arterial wall cells appeared viable in terms of dye exclusion (Moss and Benditt, 1975).

E. Media and Serum Requirements

Human SMCs grow in the same types of media and at the same serum concentrations that have been used successfully for human and animal fibroblasts. We have found that Eagle's minimum essential medium (MEM) with added glutamine and sodium bicarbonate works well with fetal aortic SMCs in 5% carbon dioxide at 37°C; Dulbecco–Vogt medium is also suitable. Fetal calf serum (FCS) (5–20%) is satisfactory for producing good cell growth in mass cultures. Dilute plating of SMCs results in higher plating (cloning) efficiency if conditioned medium is used as 20% of the total medium. Conditioned medium generally consists of medium which has been removed from a near-confluent culture after 48 hours. It is frozen and thawed to kill any viable free-floating cells and filtered through a 0.45 μm filter to be sterilized and clarified. The product is then stored frozen to preserve whatever active factors are present and necessary for improving cloning efficiency.

Ross *et al.* (1974) have reported that certain factors derived from platelets and present in whole serum, but not in plasma-derived serum, are necessary for the growth of monkey arterial SMCs. Other investigations have confirmed this finding and extended it by showing that platelet factor promotes the growth of several cell types other than SMCs (Chamley *et al.*, 1977; Kohler and Lipton, 1974; Gospodarowicz *et al.*, 1975).

F. Morphological Characteristics

Human aortic SMCs exhibit a characteristic growth pattern similar to that of animal SMCs *in vitro* (Chamley and Campbell, 1975). When cultures are dispersed with EDTA–trypsin and evenly plated into a new culture vessel, they appear flat and broad, having a rectangular rather than a spindle shape. As they migrate and multiply, they become arranged in long ridges consisting of cells piled on top of each other; at the base of the ridges cells extend out to touch adjoining piles of cells. This growth pattern, seen mostly in near-confluent cultures, gives the appearance of "wind rows" or "hills and valleys" of cells with unpopulated areas of culture dish remaining between the rows (Fig. 2).

An interesting feature of cells obtained from human fetal aortas is the regular appearance of large amounts of amorphous material in the culture medium, which occurs within 24 hours of changing the medium. Human fibroblasts either do not or only sparingly exhibit this phenomenon. Similar behavior has been observed with an abdominal aortic SMC culture from a 7-year-old boy (G. E. Striker, personal communication) and was described for SMC cultures of human umbilical cord by Gimbrone and Cotran (1975). This also has been observed with bovine aortic SMCs, but not with endothelial cells derived from bovine aorta (S. M. Schwartz, personal communication).

SMCs from human fetal aortas undergo senescent changes and cease dividing after 20 (or more) passage generations. In our laboratory the changes observed were as follows. There was a shift in morphology from a small spindle shape in those cells that made up the large, rapidly growing clones to large, broad cells present in small clones which had stopped dividing. The large cells appeared similar in morphology to large, senescent cells previously reported (Moss and Benditt, 1975). Clones consisting of small spindle cells increased in size and cell number many times over a short period of time. It was not uncommon to see mixed cell colonies or both colony types on the same plate (Fig. 3).

G.   Chemical Functions

SMCs of human arteries have been found to synthesize and secrete collagens of several types including types I, III, IV, and A α B (Layman et al., 1977; Rhodes and Miller, 1978; P. K. Killen and G. E. Striker, personal communication).

We have found that the fetal human arterial wall SMC has a mixed-function oxygenase system capable of metabolizing aryl hydrocarbons and that this system is inducible (Bond et al., 1979). Cholesterol synthesis, glycosaminoglycan synthesis, and other activities of human cells undoubtedly are present, as they are in subhuman primates (Chamley-Campbell et al., 1979).

H.   Age of Donor and Site of Origin of Tissue

Current studies in our laboratory, designed to obtain SMCs from human aorta, have met with both success and failure depending on the source of material and age of the donor. (All material was obtained with informed consent.) The best material came from fetuses up to 24 hours following spontaneous or induced abortion. When possible, thoracic and abdominal aortas were kept separate. The thoracic portion of the aorta consisted of the segment of vessel between the aortic

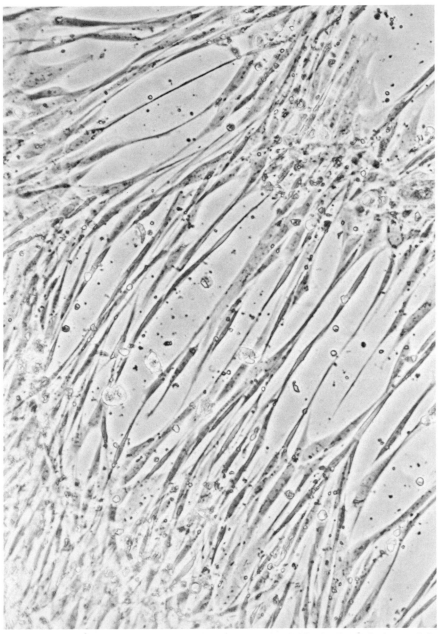

FIG. 2.   Nearly confluent culture of human embryonic aorta SMCs showing their characteristic growth pattern. Cells pile up and produce ridges of cells, while adjacent areas are nearly devoid of cells. Note also the excessive amount of "debris" floating in the culture medium. Culture medium: Eagle's MEM plus 10% FCS. Passage generation: fifth. Cell source: Thoracic aorta. Magnification: ×100.

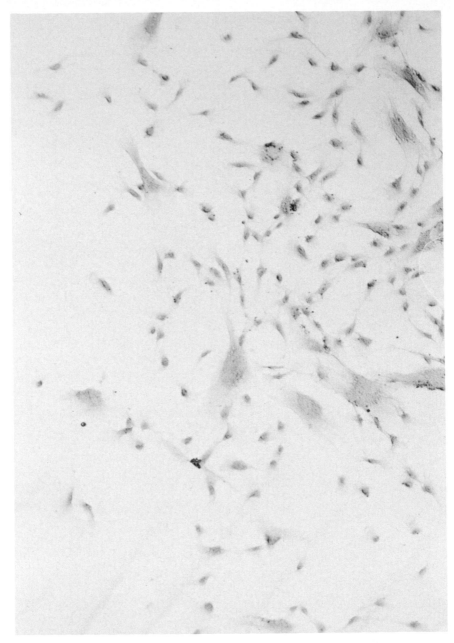

Fig. 3. A mixed cell colony showing both large, "senesced" SMCs and small, rapidly dividing SMCs. Colonies consisting of only the large cell type are small and contain relatively few cells, none of which are dividing. Colonies consisting of predominately or only small cells are large and contain thousands of cells, many of which are actively dividing. This phenomenon has been seen only in very-high-passage normal cell cultures.

arch and the lowest pair of intercostal arteries, and the abdominal portion was the segment obtained from between the celiac and common iliac arteries. The segment of aorta anterior to the diaphragm was removed and stored separately from the portion posterior to the diaphragm. The tissue was removed as aseptically as possible and placed in MEM plus 10% FCS containing antibiotics. Once in the growth medium, it was found that the aortic segments could survive as long as 48 hours if coarsely chopped and stored at 4°C. These segments were ultimately cultured as attached explants with MEM plus 10% FCS as the growth medium. Between 35 and 45% of the explant fragments produced good cell growth by 10 days in culture. A systematic attempt to determine the maximum number of doublings possible with these cells is in progress. However, more than 20 *in vitro* doublings have been observed in several cases.

We successfully grew SMCs from 23 fetuses, including one having Down's syndrome. The population doubling time for these cells ranged from 36 to 72 hours, somewhat longer than that observed for newborn foreskin fibroblasts. Cloning was carried out in MEM plus 15% FCS and 20% conditioned medium. As is well known from studies with human fibroblasts, cloning efficiency varied with the serum lot used (Table I). Based on the serum lot and the presence or absence of conditioned medium, the cloning efficiency for fetal aortic SMCs ranged from 5 to 25%. As previously noted, we found that fetal aortic tissue would remain viable for up to 48 hours prior to culturing when kept refrigerated in MEM-10 with HEPES buffer (pH 7.2).

In our recent experience with material from numerous autopsy cases of both children and adults we have had one successful culture of artery wall cells from the abdominal aorta of an 18-year-old female taken 12 hours postmortem. Successful outgrowths of arterial SMCs have been regularly obtained from perfused artery maintained for kidney transplant purposes (G. E. Striker, personal communication).

The "plugs" or "buttons" removed from the aorta during coronary artery bypass graft surgery seemed a good potential source of human arterial wall SMCs. These specimens were collected directly at operation into sterile culture media, transported to our cell culture facility, and minced into 1-mm fragments for attached explant culture. Some of the fragments were treated with collagenase prior to culturing, some were allowed to adhere to the bottom of the dishes and then covered with culture medium, and others were placed under coverslips prior to incubation in the medium.

Explants were cultured in several different media, each containing 10–15% FCS. Eagle's MEM, RPMI-1640, L-15, and Dulbecco–Vogt were all tested. In addition, some explants were provided with conditioned medium (CM) which consisted of medium plus serum that had supported the growth of near-confluent human fibroblasts or human embryo aortic SMCs for 48–72 hours.

In one set of experiments a total of 364 explants from 5 subjects, representing

TABLE I

Cloning Efficiency of Human Fetal Aortic Smooth Muscle Cells with and without 20% Conditioned Medium[a]

Type of cell	A	A + CM	Increase (%)	B	B + CM	Increase (%)	C	C + CM	Increase (%)	D	D + CM	Increase (%)	E	E + CM	Increase (%)
Thoracic	4.2	11.2	62	2.8	10.2	72	0.8	3.8	79	1.0	3.2	69	1.8	3.4	47
Abdominal	5.2	9.0	42	1.6	5.8	72	0.8	6.6	88	0.8	2.8	71	0.8	5.2	85

[a] Mean increase in cloning efficiency with CM: 66% for thoracic and 71% for abdominal SMCs.

13 separate pieces of freshly obtained aortic tissue, yielded 3 fragments which produced a few migrating cells. None of these explants produced cultures of dividing SMCs after 5 weeks in culture at 37°C and in a 5% carbon dioxide atmosphere. This experiment was repeated several times with essentially the same result: Less than 5% of tissue fragments produced cells, and none of the cells continued to proliferate.

With human femoral artery it has been possible with reasonable frequency to obtain cellular outgrowths that have proliferated *in vitro* (Moss and Benditt, 1975). Femoral artery was obtained directly from surgical specimens and immediately minced into fragments for organ culture, or was obtained postmortem. Specimens obtained up to 12 hours postmortem were found to be viable in this series of investigations. Studies made on individuals ranging in age from 46 to 81 years produced femoral artery cells capable of up to 16 population doublings. There was in this series no obvious correlation with the age of the donor. Bierman (1978) reported that cells from subjects in this age range only doubled 6-8 times and that the maximum number of doublings observed was about 10-14 times in cells derived from newborns. His data showed great variability in the number of doublings, as did ours (Moss and Benditt, 1975), suggesting that more careful studies on factors affecting the variability of life span under culture conditions need to be performed.

I.   Human Aortic Organ Culture

Studies on the maintenance in culture of the buttons of ascending human aorta removed during the performance of coronary bypass surgery were made as an initial step in the development of metabolic studies. It was observed that FCS (5-10%) in the medium stimulated the appearance of nucleoli in many cells of human aortic fragments and that with this appeared substantial necrosis of many cells in the tissue fragments. However, in the absence of serum, cultivation of the tissue fragments in basal medium resulted in good maintenance of morphological integrity for 1-3 weeks. Of the three types of medium used, medium M199 resulted in the best tissue matinenance; the other two, Dulbecco–Vogt and L-15, appeared to be less satisfactory. Addition of [³H]uridine to the medium after 1 week in serum-free cultures revealed uptake of the labeled RNA precursor by autoradiography in essentially all the surviving cells after a 6-hour exposure to uridine. Exposure of tissue explants to labeled proline in serum-free medium for 6 hours also indicated that the bulk of the cells were able to take up and to incorporate the amino acid into protein.

J.   Other Observations on Cultured Smooth Muscle Cells

Bierman and Albers (1975) and Albers and Bierman (1976) have reported that human arterial SMCs in culture preferentially take up lipid-rich lipoproteins

(VLDL and LDL) and that hypoxia of the cultures does not alter the binding of LDL but does increase its net uptake and decrease its degradation.

Also, Stout *et al.* (1975) reported that primate arterial SMCs in culture exhibited increased growth rates when exposed to insulin. They noted, however, that this growth-stimulating feature disappeared when the cultures had been passed eight or nine times. They could not see any morphological alteration as a result of insulin exposure.

K.   Mutation in Human Aortic Smooth Muscle Cells

In experiments in our laboratory designed to examine the susceptibility of human arterial wall SMCs to mutagens we have observed that SMCs from fetal aorta respond similarly to human fibroblasts and other mammalian cells (De-Mars, 1974; Chu and Malling, 1968). With N-methyl-N-nitro-N-nitrosoguanidine (MNNG), a direct mutagen, and exposure for 2 hours to a concentration of 0.2 μg/ml, we have obtained clones that are resistant to ouabain at a concentration 100 times that which is lethal to the wild-type precursor vascular SMCs. Evidence from one case suggests that abdominal SMCs may be 10 times more susceptible to induced mutation, in our system, than thoracic SMCs. The spontaneous mutation frequency for ouabain resistance in wild-type cells was less than 1×10^{-8}, while the frequency for abdominal cells was 1.6×10^{-6} and 1.5×10^{-7} for thoracic cells.

III.   Summary

Adult human aortic SMCs are more difficult to cultivate *in vitro* using conventional cell culture procedures than human fibroblasts. The age of the donor and the site of origin, as well as the nature of the medium, condition the result.

Human fetal aorta readily produces SMCs from explants when grown in 10% FCS and conventional cell culture media. Some properties of human fetal aortic SMCs observed *in vitro* are: (1) Near-confluent SMC grow in ridges or "wind rows" of piled up cells; (2) their doubling time is 36–72 hours; (3) dilute cloning efficiency with 20% conditioned medium is 5–25%; (4) fetal human SMCs have a mixed-function oxygenase system; (5) and mutations to ouabain resistance can be induced in human fetal aortic SMCs by exposure to the direct mutagen MNNG, and the frequency is similar to that found with human fibroblasts.

In culture, buttons of adult thoracic aorta removed from surgery have not yielded viable cell outgrowths. In organ culture the tissues exhibit extensive cell necrosis in serum-containing, but not in serum-free media. The nature of the toxic effect needs further investigation.

Appendix

A. Smooth Muscle Cell Explants from Human Arteries

1. Obtain arteries as source of SMCs.
2. Remove the artery tissue as sterile as possible to a container with culture medium. (Culture medium for explant procedure should contain antibiotics.)
3. Wash the tissue several times with culture medium.
4. Using sterile instruments cut 2-mm cross sections of artery. Remove all connective tissue that would be a source of contaminating fibroblasts and epithelial cells. Put the cleaned strips of SMC tissue into a dish of fresh culture medium.
5. With scissors or scalpels cut the SMC tissue into smaller pieces—approximately 1 mm^3.
6. With a sterile Pasteur pipet transfer 1-mm^3 pieces of tissue to the culture dish in which they are to grow. Use the Pasteur pipet to distribute the pieces evenly (optimum density, four per square centimeter).
7. After the pieces of artery are evenly spread, use the Pasteur pipet to remove as much medium as possible from the culture dish. Put the dish into a humidified 37°C incubator 2–4 hours to allow the explants to stick.
8. Gently add enough medium to cover the explants without dislodging them.
9. Change the medium weekly. Outgrowth should be apparent by 10–30 days.

B. Cloning Human Aortic Smooth Muscle Cells

1. Trypsinize (0.25% trypsin–0.2% EDTA) a mass culture, making certain that the resulting cell suspension does not contain clumps of cells.
2. Count the cells and prepare a dilution that will yield 100 cells per 10 ml medium.
3. Dispense 10 ml of cell suspension per 100-mm plastic tissue culture dish. Allow cells to settle and attach 1–2 hours without disturbing and then move plates to a 37°C incubator.
4. Feed plates weekly with fresh medium for 3–4 weeks and inspect for clones.
5. When clones are 1–2 mm in diameter, fix and stain for counting, or remove them to individual vessels. Wash the plate two times with Ca^{2+}- and Mg^{2+}-free PBS. Put a sterile cloning collar with sterile vacuum grease on the bottom end over a clone. Press the collar down so the vacuum grease seals the collar to the plate. With a Pasteur pipet put trypsin–EDTA into the cloning collar. When the cells of the clone are rounded up, gently pipet the trypsin up and down with a Pasteur pipet.

6. Place the cell suspension in a small culture dish and add the culture medium.

7. Continue expanding the culture by 2:1 splits as the cells grow.

REFERENCES

Albers, J. J., and Bierman, E. L. (1976). *Biochim. Biophys. Acta* **424**, 422–429.
Bierman, E. L. (1978). *Fed. Proc., Am. Soc. Exp. Biol.* **37**, 2832–2836.
Bierman, E. L., and Albers, J. J. (1975). *Biochim. Biophys. Acta* **388**, 198–202.
Bond, J. A., Kocan, R. M., Benditt, E. P., and Juchan, M. R. (1979). *Life Sci.* **25**, 425–430.
Chamley, J. H., and Campbell, G. R. (1975). *Cytobiologie* **11**, 358–365.
Chamley, J. H., Campbell, G. R., McConnell, J. D., and Groschel-Stewart, U. (1977). *Cell Tissue Res.* **177**, 503–522.
Chamley-Campbell, J. H., Campbell, G. R., and Ross, R. (1979). *Physiol. Rev.* **59**, 1–61.
Chu, E. H. Y., and Malling, H. V. (1968). *Proc. Natl. Acad. Sci. U.S.A.* **61**, 1306–1312.
DeMars, R. (1974). *Mutat. Res.* **24**, 335–364.
Fischer-Dzoga, K., Jones, R. M., Vesselinovitch, D., and Wissler, R. W. (1973). *Exp. Mol. Pathol.* **18**, 162–176.
Gimbrone, M. A., Jr., and Cotran, R. S. (1975). *Lab. Invest.* **33**, 16–27.
Gospodarowicz, D., Green, G., and Moran, J. (1975). *Biochem. Biophys. Res. Commun.* **65**, 779–787.
Jarmolych, J., Daoud, A. S., Landau, J., Fritz, K. E., and McElvene, E. (1968). *Exp. Mol. Pathol.* **9**, 171–188.
Kohler, N., and Lipton, A. (1974). *Exp. Cell Res.* **87**, 297–301.
Layman, D. L., Epstein, E. H., Dodson, R. F., and Titus, J. L. (1977). *Proc. Natl. Acad. Sci. U.S.A.* **74**, 671.
Mark, G. E., Chamley, J. H., and Burnstock, G. (1973). *Dev. Biol.* **32**, 194–200.
May, J. F., Paule, W. J., Rounds, D. E., Blankenhorn, D. H., and Zemplenyi, T. (1975). *Virchows Arch. B* **18**, 205–211.
Moss, N. S., and Benditt, E. P. (1975). *Am. J. Pathol.* **78**, 175–186.
Rhodes, R. K., and Miller, E. J. (1978). *Biochemistry* **17**, 3442.
Ross, R. (1971). *J. Cell Biol.* **50**, 172–186.
Ross, R., Glomset, J., Kariya, B., and Harker, L. (1974). *Proc. Natl. Acad. Sci. U.S.A.* **71**, 1207–1210.
Rossi, G. L., Alroy, J., and Rothenmund, S. (1973). *Virchows Arch. B* **12**, 133–144.
Stout, R. W., Bierman, E. L., and Ross, R. (1975). *Circ. Res.* **36**, 319–327.

Chapter 11

The Fetal Mouse Heart in Organ Culture: Maintenance of the Differentiated State[1]

JOANNE S. INGWALL

Department of Medicine,
Harvard Medical School and Peter Bent Brigham Hospital,
Boston, Massachusetts

WILLIAM R. ROESKE

Department of Internal Medicine,
University of Arizona, School of Medicine
Tuscon, Arizona

AND

KERN WILDENTHAL

Pauline and Adolph Weinberger Laboratory for Cardiopulmonary Research,
Departments of Physiology and Internal Medicine,
University of Texas Health Science Center at Dallas,
Dallas, Texas

[1]This work was supported in part by grants from the Massachusetts and San Diego County affiliates of the American Heart Association and by the NIH Specialized Center of Research on Ischemic Heart Disease HL-17682, HL-22290, and HL-21751. Joanne S. Ingwall is an established investigator of the American Heart Association.

I. Introduction

Myocardial metabolism, structure, and function have been studied in many *in vitro* preparations, including isolated perfused hearts, papillary muscle strips, and heart cells in monolayer culture. Intact heart preparations offer the advantage of studying the whole functioning organ but suffer from the disadvantages of being stable for only brief periods of time and of being difficult to control. In contrast, preparations of spontaneously beating myocardial cells in monolayer culture offer the advantages of ease and reproducibility of obtaining many replicate heart preparations, the ability to manipulate experimental conditions, and long-term stability; but extrapolation of the responses of isolated cells to the intact heart is limited. There is a culture model that combines the advantages of working with a spontaneously beating intact heart with the advantages of a culture model: the intact beating fetal mouse heart in organ culture. Hearts from 15- to 20-day-old fetal mice (term is approximately 21–22 days) are small enough (0.5–2 mm in diameter) to be adequately supplied with oxygen and nutrients by diffusion and yet are sufficiently well developed to exhibit many characteristics of the fully differentiated adult heart. These include spontaneously rhythmic contractions, appropriate responsiveness to neurotransmitters, and synthesis of organ-specific substances. As with any organ culture preparation, however, a slow, progressive loss of some characteristics of differentiation occurs, and catabolic processes become dominant. In this chapter we describe (1) some of the morphological, functional, and biochemical measurements that have been made on hearts weighing only 2–5 mg, (2) the influence of fetal age and time in culture on the biochemical expression of differentiation, and (3) the effects of supplying hearts with various hormones and other agents on net cardiac protein balance and on the synthesis of organ-specific proteins.

II. Methods and Materials

A. Organ Culture Preparation

Pregnant mice have 6–14 fetuses, which provide 6–14 matched littermate hearts suitable for culturing. With the following procedure, hearts isolated from pregnant mice during the last trimester of gestation can be successfully maintained in culture for 4–8 days.

Immediately following cervical dislocation and hysterectomy of the pregnant mouse, fetuses are removed from the uterus aseptically and decapitated to permit extrusion of blood from the heart. After removing the lower half of the body at the level of the diaphragm, the chest cavity is cut open to expose the heart. In order to avoid damaging the heart, the heart is not directly manipulated with forceps but rather removed via the thymus. Hearts are placed in culture dishes containing 1–2 ml of medium and, with the aid of a dissecting microscope, dissected free of the thymus, pericardium, and vessels. Hearts are then transferred to 0.5-mm stainless steel grids in organ culture dishes using a wide-orifice pipet and supplied with sufficient culture medium so that they are cultured at a liquid–gas interface. This configuration is shown in Fig. 1. Approximately 400 μl of culture medium [minimum essential medium (MEM), medium 199, or Earle's salt solution] fills commercial culture dishes (such as those made by Falcon); 0.75–1.5 ml of medium is needed for the organ culture dishes described by Fell and colleagues (1968). Organ culture dishes are kept in airtight culture jars in an atmosphere of 95% oxygen and 5% carbon dioxide at 37°C. The minimum percentage of oxygen needed for viability is ~30%. Further details of this culture method have been described by Wildenthal (1971).

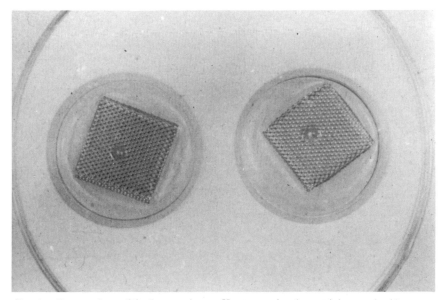

FIG. 1.   Organ culture of fetal mouse hearts. Hearts are placed on stainless steel grids at gas–liquid interface.

B.   Morphological Analyses

For both light and electron miscroscopic evaluation, hearts are immersed in 3% gluteraldehyde in 0.1 M cacodylate buffer at pH 7.4 and allowed to fix for 3 hours. They are then washed in cacodylate buffer, postfixed with 1% osmic acid, dehydrated through graded alcohols and propylene oxide, and embedded in Epon. For light microscopy, sections are cut to approximately 1 μm and stained with toluidine blue. For electron microscopy, ultrathin sections are stained with alcoholic uranyl acetate and lead citrate.

C.   Measurement of Mechanical Performance

The sonometric technique developed for large animal hearts by Théroux *et al.* (1974) has been adapted to 3-mg fetal mouse hearts maintained in culture (Roeske *et al.*, 1976). In this experiment, the transit time of ultrasound between two piezoelectric crystals is measured. A difference in the transit time corresponds to a difference in the distance between the two crystals. The arrangement of the crystals and the heart in the culture dish is shown in Fig. 2. The receiver crystal is made by soldering single strands of 25-strand stainless steel wire to a

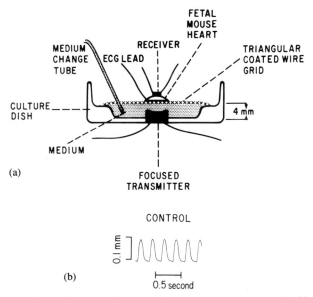

FIG. 2.   (a) A culture dish adapted for sonomicrometer measurement. A 5-mHz focused transmitter transmits a signal to the superior surface of the ventricle. (b) An example of a tracing monitoring change in thickness of the heart and frequency of contraction of a 17-day-old heart cultured for 1 day.

1-mg crystal; it adheres to the superior aspect of the ventricle by surface tension. The transmitter crystal (and lens) is glued to the bottom of the dish and functions to focus the sound beam at the level of the cultured heart. By triggering on single cycles of received sound, differences in the distance between the two crystals as small as 2 μm can be detected. From these measurements, the frequency of contraction can be precisely measured. In addition, the change in heart thickness during a beat provides a measure of the contractile state of the heart.

Other methods that have been employed to measure mechanical performance in organ-cultured hearts include the use of capacitance-sensitive devices (Wildenthal et al., 1973) and laser beams (Mauer et al., 1979).

D.   Biochemical Analyses

1.   CREATINE KINASE AND LACTIC DEHYDROGENASE ACTIVITIES

For the measurement of creatine kinase (CPK) and lactic dehydrogenase (LDH) activities and CPK isozyme composition, hearts are homogenized in 0.1 M phosphate buffer, pH 7.4, with 1 mM EDTA and 1 mM β-mercaptoethanol. CPK activity is measured using the methods of Rosalki (1967) and of Szasz et al. (1970). Quantification of CPK isozymes is made by measuring the fluorescence of NADPH generated by the isozymes separated by cellulose acetate strip electrophoresis as described by Hall and DeLuca (1976). LDH activity is measured in the direction of lactate production (Bernstein and Everse, 1975).

2.   LYSOSOMAL ENZYME ACTIVITIES

For measurements of lysosomal enzyme activities, hearts are homogenized in a 0.1% solution of Triton X-100 with a Polytron homogenizer to disrupt cells and organelles maximally. Activities of cathepsin D, β-acetylglucosaminidase, and acid phosphatase are assayed using a modification of the methods of Barret (1972) as previously described (Wildenthal et al., 1975).

3.   ATP AND CREATINE PHOSPHATE CONCENTRATIONS

For analyses of ATP, creatine phosphate, and creatine concentrations, hearts are rapidly frozen in liquid nitrogen. Following homogenization in 0.4 M perchloric acid, extracts are neutralized with potassium phosphate, pH 6.5, and clarified by centrifugation. ATP and creatine phosphate contents are measured using the methods of Lowry and Passoneau (1972). Total creatine (i.e., free creatine plus creatine phosphate) is measured using the fluorometric assay of Kammermeier (1973).

4. OTHER ANALYSES

In order to determine changes in cardiac mass, the protein (Lowry *et al.*, 1951), DNA (Hinegardner, 1971), and water contents of hearts are measured. Measurement of total tissue water is made by drying ⩾20 mg wet weight of tissue to constant weight. Extracellular space is determined as sorbitol space using [^{14}C]sorbitol as previously described (Griffin and Wildenthal, 1978a).

E.   Statistical Methods

For most biochemical analyses and for morphological and functional measurements, single hearts are used; however, for measurements of ATP, creatine phosphate, and DNA contents, a minimum of eight hearts for each analysis is necessary to obtain reliable results.

Results obtained for individual hearts within a single litter or from different litters of the same age are highly reproducible (standard deviation <10%). Whenever possible, therefore, comparisons are made for matched littermate hearts using Student's *t*-test for paired data. For all other comparisons, Student's *t*-test for unpaired data is used. Results are given as the mean ± 1 SEM, and *n* is the number of hearts tested.

III.   Results and Discussion

A.   Morphology

1. CHANGES IN CARDIAC MASS

Hearts from fetuses 16 days old or older have well-formed atria and ventricles. Supplied with a chemically defined culture medium, hearts of all ages survive in culture for 4–8 days, but lose myocardial mass at a rate of 10–20% per day. To illustrate this process, changes in water, protein, and DNA contents and in total cardiac mass with time in culture for hearts from 17-day-old fetuses are shown in Table I. Protein and water contents decrease proportionally with decrease in mass, but the amount of DNA per heart remains the same. These results suggest that the number of cells per heart remains the same with time in culture, but that there is a progressive decrease in the size of the cells. It is possible, however, that a small number of cells die each day and that the cellular debris consists in part of

TABLE I

Changes in Cardiac Mass with Time in Culture for 17-Day-Old Fetal Mouse Hearts
Cultured in MEM[a]

Parameter	Freshly explanted		Time in culture			
			Day 1		Day 2	
Milligrams wet weight per heart	2.40 ± 0.08	(4)	1.93 ± 0.03	(7)	1.69 ± 0.09	(4)
Milligrams water per milligram wet weight			0.825 ± 0.004	(9)	0.818 ± 0.010	(3)
Extracellular space (mg water/mg wet weight)[b]			0.326 ± 0.007	(18)	0.327 + 0.017	(9)
Intracellular space (mg water/mg wet weight)			0.499		0.491	
Protein content[c]						
Milligrams protein per milligram wet weight	0.115 ± 0.007	(4)	0.113 ± 0.005	(16)	0.120 ± 0.004	(9)
DNA content[d]						
Milligrams DNA per heart	8.9 ± 0.5	(4)	10.3 ± 0.4	(6)	10.0 ± 0.8	(4)
Milligrams DNA per milligram wet weight	3.7 ± 0.3	(4)	5.3 ± 0.2	(6)	5.7 ± 0.2	(4)

[a] Values in parentheses refer to the number of analyses performed.
[b] Measured as sorbitol space (Griffin and Wildenthal, 1978a).
[c] Measured using the method of Lowry *et al.* (1951).
[d] Measured using the method of Hinegardner (1971).

DNA-containing nuclei. Morphological analysis suggests that both possibilities can occur.

2.   Light Microscopy

In preliminary experiments performed in collaboration with R. A. Kloner of Peter Bent Brigham Hospital, hearts were maintained in culture for 1 or 2 days, fixed, and sliced along their major axis for light microscopic analyses (Kloner and Ingwall, 1980). Hearts from 17-day-old fetuses showed a uniform population of healthy cells; however, larger hearts from older fetuses and hearts deliberately deprived of oxygen had a core of necrotic cells containing primarily pycnotic nuclei. The larger the heart or the longer the period of hypoxic insult, the larger the necrotic core. Thus, cultured hearts are composed of viable, differentiated cells surrounding a core of cell debris. This core is virtually in hearts from 17- to 18-day-old fetuses but increases in size with increasing size of the heart. Therefore, loss of cardiac mass in large hearts probably results from both cell death and atrophy of viable cells, while loss of cardiac mass from small hearts reflects primarily cell atrophy.

3. ELECTRON MICROSCOPY

Ultrastructural analysis provides important information about cell viability and the extent of differentiation. Features useful in assessing cellular viability include (1) degree of cell swelling, (2) dispersal of nuclear chromatin, (3) integrity of the mitochondria (particularly of the cristae), and (4) integrity of the myofibrils (particularly the interfilament spacing and density of the Z band). The microgaph shown in Fig. 3 permits examination of three myocytes in an 18-day-old fetal heart cultured for 1 day. The nuclei have well-dispersed chromatin, the mitochondria have a normal appearance, and the myofibrils, while low in density compared to those in the adult myocardium, are well formed with sharp Z bands. These are characteristics of viable myocytes capable of contraction.

B. Mechanical Performance

The nature of the atrial-ventricular activation sequence and frequency of contraction of cultured hearts can be identified simply by visual inspection using a light microscope. Hearts beat spontaneously and rhythmically throughout the culture period. The transition to death occurs rapidly.

In order to quantitate mechanical activity and function, the sonometric technique, which has proved to be useful in measuring changes in tissue dimensions in large animal hearts, has been applied to the cultured heart. In such experiments, the transit time of ultrasound is measured between two piezoelectric crystals located on either side of the heart. In diastole, the cultured heart is relatively flat. During systole, the heart actively contracts and assumes a more globular shape. The increased distance between the crystals during systole provides an index of the change in heart thickness associated with contraction and allows accurate determination of the frequency of contraction. An example of a tracing monitoring the change in thickness and frequency of contraction using this technique is shown in Fig. 2. In this example, the heart rate was 261 beats per minute, and the change in heart thickness during contraction was 0.1 mm.

C. Biochemical Characteristics

1. INFLUENCE OF FETAL AGE

Hearts obtained from 14- to 21-day-old fetuses are progressively more differentiated. Substrate utilization assumes a more adultlike pattern with increasing age (Wildenthal, 1973a), and responsiveness to cardioactive drugs becomes more mature (Wildenthal, 1973b; Wildenthal *et al.*, 1976a). Increasing differentiation with age is reflected by changes in the specific activities of several

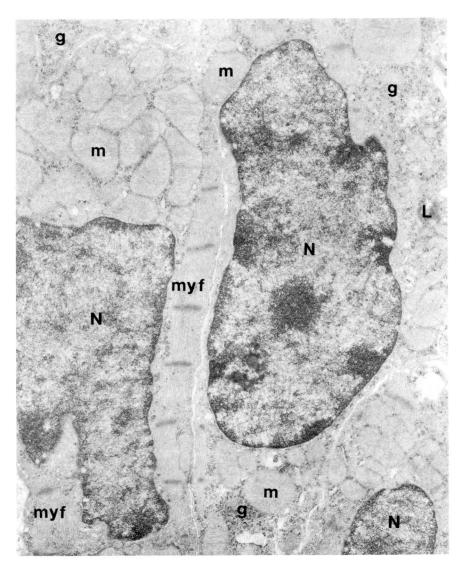

FIG. 3.   Electron micrograph of an 18-day fetal house heart maintained in organ cultures for 1 day. Three myocytes are pictured, displaying nuclei (N), mitochondria (m), glycogen granules (g), lysosomes (L), and myofibrils (myf). Magnification: ×18,000. The micrograph was graciously provided by Dr. Robert S. Decker and Dr. Robert M. Ridout.

cardiac enzymes. For example, the activity of the muscle-specific protein CPK monotonically increases from 50 ± 5 mIU/mg wet weight in 0.5-mg hearts isolated from 14-day fetuses to 450 ± 20 mIU/mg wet weight in 8-mg hearts from 21-day fetuses (Fig. 4). The increase in accumulation of CPK is due primarily to an increased accumulation of the muscle-specific isozyme MM-CPK. In hearts from 14-day fetuses, the specific activity of MM-CPK is barely measurable (~ 1 mIU/mg wet weight), but in hearts isolated from fetuses at term the specific activity of MM-CPK is 325 ± 20 mIU/mg wet weight. In contrast to these impressive changes in the accumulation of CPK, the specific activity of the glycolytic enzyme LDH does not change during this time of development (Fig. 4). These results suggest that different classes of proteins accumulate at different rates during development. The observation that the specific activity (and presumably tissue concentration) of MM-CPK increases more than two orders of magnitude during the last trimester of development (only 7 days) emphasizes the dynamic nature of developing tissue and thus the need for designing experimental protocols using hearts of the same size.

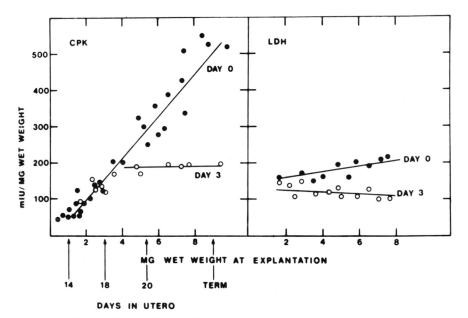

Fig. 4. The specific activities of CPK (left) and LDH (right) in freshly explanted fetal mouse hearts (day 0, solid circles) and after 3 days in culture (day 3, open circles). CPK activity was measured using the coupled enzyme method of Rosalki (1967) and LDH activity using the method of Bernstein and Everse (1975). Each entry represents the mean of enzyme activities obtained from 2–20 hearts and is accurate within 10%. From Recent Advances in Studies on Cardiac Structure and Metabolism, Volume 12. Edited by Tachio Kobayashi, Yoshio Ito, and George Rona. Copyright © 1978 by University Park Press, Baltimore, Maryland. Reprinted with permission.

2.   INFLUENCE OF TIME IN CULTURE

a.   Net Protein Balance.   The differentiated state is maintained in culture, but degradation exceeds synthesis (Wildenthal, 1972; Wildenthal *et al.,* 1976b). Experiments measuring rates of synthesis and degradation of total cardiac protein at different times in culture indicate that the balance between synthesis and degradation remains relatively constant once the hearts are cultured. The rate of protein synthesis is the same in 19- to 20-day-old hearts cultured for 1 or 3 days (Ingwall and Wildenthal, 1978). The rate of protein degradation, as measured by the release of amino acids from hearts into the culture medium, is similar in hearts cultured for 1 or 2 days (Wildenthal and Griffin, 1976). The rate of loss of radiolabeled protein in hearts cultured for 1 or 2 days is also the same (Wildenthal *et al.,* 1978). These results suggest that the changes in the rates of total cardiac protein synthesis and degradation occur soon after the heart is explanted but are maintained relatively constant thereafter.

Not all proteins or classes of proteins, however, are synthesized and degraded at the same rate observed for total cardiac protein. The rate of synthesis of myosin heavy chain, the major myofibrillar protein, in 19- to 20-day-old hearts cultured for 3 days is approximately two-thirds the rate observed in littermate hearts cultured for 1 day (Ingwall and Widlenthal, 1978). The progressive loss of CPK and LDH activity with time in culture shows that the accumulation as well as the synthesis rate of specific proteins decreases with time in culture. By day 3 in culture, the specific activity of CPK has decreased to $\sim$200 mIU/mg wet weight for all hearts weighing $\geqslant$4 mg (Fig. 4). Changes in the specific activity of MM-CPK with time in culture and with increasing size of tne heart are similar to the changes observed for total CPK activity.

These results suggest that the concentrations of several cardiac muscle proteins (myosin, CPK, and LDH) decrease with time in culture, even though the amount of total cardiac protein per unit mass (see Table I) remains the same. Moreover, the changes are progressively greater as the age and size of the heart increases.

Since cultured hearts are in negative protein balance, it might be expected that proteins that play a role in degradation would accumulate with time in culture. Changes in the specific activities of several lysosomal hydrolytic enzymes as a function of time in culture are shown in Fig. 5. The specific activities of acid phosphatase, cathepsin D, and β-acetylglucosaminidase all increase significantly with time in culture. This accumulation of lysosomal enzymes is in sharp contrast to the decrease in concentration of myosin, CPK, and LDH and suggests that the concentration of proteins within a class (muscle-specific proteins on the one hand and lysosomal enzymes on the other) may be subject to coordiante control.

b.   ATP Levels.   Just as the concentration of muscle-specific proteins decreases with time in culture, the ATP concentration also decreases. Compared to values found for freshly explanted hearts, ATP levels fall approximately 30% in

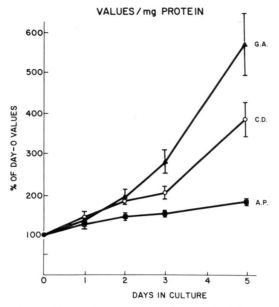

FIG. 5. Change in specific activities of acid phosphatase (A.P.) cathepsin D (C.D.), and β-acetylglucosaminidase (G.A.) in 4.8-mg hearts with time in culture. Values are expressed as percentage of values found for freshly explanted hearts: acid phosphatase, 275 ± 8 nmoles nitrophenol/hour/mg protein ($n = 4$); cathepsin D, 44 ± 1 μg tyrosine/hour/mg protein ($n = 4$); β-acetylglucosaminidase, 256 ± 15 nmoles nitrophenol/hour/mg protein ($n = 4$). Assays were performed as described by Wildenthal *et al.* (1975).

2 days: from 34.8 ± 1.4 ($n = 24$) to 23.8 ± 0.9 ($n = 15$) nmoles/mg protein. In spite of this decrease in concentration, however, the adenylate energy charge (defined by Atkinson as ATP + ½ADP–ATP + ADP + AMP) remains about the same: 0.91 ± 0.01 ($n = 17$) for freshly explanted hearts and 0.88 ± 0.01 ($n = 8$) for hearts cultured for 2 days. This indicates that the availability of higher-energy phosphate bonds for energy-requiring processes is similar in freshly explanted and cultured hearts.

Experiments measuring the rate of loss of radiolabeled nucleosides from cultured hearts show that freely diffusible purines such as inosine and hypoxanthine are slowly but progressively lost ($\leq 1\%$ per hour) from hearts into the culture medium (Kaufman *et al.*, 1977). The loss of purine nucleotide precursors may explain the decrease in ATP levels observed.

3. Effects of Supplying Hormones and Other Agents

Results presented in Sections I,A and C show that maintaining fetal hearts in organ culture results in a progressive loss of cardiac mass. In Table II, a sum-

TABLE II

Summary of Effects of Various Agents on Net Protein Balance in Cultured Fetal Mouse Hearts

	Effect on protein balance	
Polypeptids and hormones		
Aprotonin	No change	Libby et al. (1979)
Glucagon	No change	Wildenthal et al. (1976b)
Growth hormone with insulin	Increased synthesis	Griffin and Wildenthal (1978b)
Growth hormone without insulin	Decreased degradation	Griffin and Wildenthal (1978b)
Hydrocortisone without insulin	Decreased synthesis, increased degradation	Griffin and Wildenthal (1978a)
Hydrocortisone with insulin	Increased synthesis, decreased degradation	Griffin and Wildenthal (1978a)
Insulin	Increased synthesis, decreased degradation	Wildenthal et al. (1976b)
Soybean trypsin inhibitor	No change	Libby et al. (1979)
Thyroxine	Increased synthesis, slightly decreased degradation	Wildenthal et al. (1979)
Low-molecular-weight substances		
Antipain	Slightly decreased degradation	Libby et al. (1979)
Chymostatin	No change	Libby et al. (1979)
Chloroquine	Decreased degradation	Wildenthal et al. (1978)
Creatine	Slightly increased muscle-specific protein synthesis	Ingwall and Wildenthal (1976)
Leupeptin	Decreased degradation	Libby et al. (1979); Wildenthal and Crie (1979)
Pepstatin	No change	Libby et al. (1979)
Sucrose	Decreased degradation	Wildenthal et al. (1978)

mary of agents tested for their ability to alter degradation or synthesis is given. In some cases, such as with insulin, the effects are general and the mechanisms involved are not well defined. However, in others, manipulations are intended to alter specific metabolic processes in the hope of improving net protein balance. Some agents that improve net protein balance do so by simultaneously increasing the rate of protein synthesis and decreasing the rate of protein degradation. It is also of interest that the action of some agents (e.g., hydrocortisone and growth hormone) is modified by the presence or absence of insulin.

Several agents known to disrupt lysosomal function or to inhibit lysosomal enzymes have been tested for their ability to retard degradative processes. Chloroquine and sucrose both cause lysosomal dysfunction and both are effective in decreasing the rate of degradation and increasing the rate of synthesis. A variety of protease inhibitors have also been tested for their ability to reduce degradation. These include a series of peptide aldehydes produced by actinomycetes, namely, antipain, leupeptin, pepstatin, and chymostatin, and the polypeptides soybean trypsin inhibitor and aprotonin. Of these, only leupetin and antipain decrease the rate of protolysis. Leupetin (30 μM) slows the rate of protein breakdown ~50% in 48 hours, and antipain (30 μM) only ~19%.

Attempts were also made to increase intracellular concentrations of ATP and creatine phosphate (and thereby prolong viability) by supplying hearts with precursors of high-energy phosphate-containing compounds. In preliminary experiments in which inosine, a nontoxic purine nucleotide precursor, was supplied to hearts, the decrease in ATP concentration with time in culture was slowed (~15% per day).

In other experiments, creatine (5 mM) was supplied. After 2 days in culture, total creatine concentration was sevenfold higher (11.8 ± 0.9, $n = 5$, to 81.7 ± 3.9 $n = 8$, nmoles/mg protein), and the creatine phosphate concentration was threefold higher (9.0 ± 0.6, $n = 5$, to 26.6 ± 2.6, $n = 7$, nmoles/mg protein) than in control cultures. The ATP concentration was unchanged (23.8 ± 0.9, $n = 15$, compared to 21.1 ± 1.4, $n = 8$, nmoles/mg protein). The effects of supplying hearts with creatine on the rates of synthesis and accumulation of several cardiac proteins have also been measured (Ingwall and Wildenthal, 1976; Ingwall, 1976) and can be summarized as follows: (1) small but significant increases (5–10%) in the rate of incorporation of labeled amino acid into total cardiac protein and in total protein content, (2) significant increases (20–40%) in the rate of myosin heavy-chain synthesis and in specific activities of total CPK and MM-CPK, and (3) no change in the activities of non-cardiac-specific proteins such as LDH, β-acetylglucosaminidase, and cathepsin D. The increases in synthesis of muscle-specific proteins and in creatine phosphate levels may be causally related.

In summary, a variety of agents have been tested for their effects on general protein balance and on specific metabolic pathways in the cultured fetal mouse heart. In addition to identifying growth-promoting agents which may be useful additives to the culture medium, results from such experiments may provide insights into regulatory mechanisms operating in the heart.

IV. Perspectives

A. Current Applications and Findings

The cultured fetal mouse heart preparation has been used to address many diverse problems in cardiology, especially in the area of metabolism. Interpretation of results can be made independently of potentially complicating neural and humoral factors and of flow-dependent changes in vascular resistance associated with more complex heart preparations. Generally the experimental protocols devised take advantage of the following features of the cultured heart preparation: a large number of matched littermate hearts, relative ease of preparing many cultures per day (60–80 is not unreasonable), ease of manipulating experimental conditions, long-term stability and, finally, the ability to correlate biochemical measurements with changes in mechanical performance and structure. Specific examples of how the cultured mouse heart has been used to define maturation of various biochemical pathways and to increase our understanding of the response of myocardial tissue to injury are given below.

1. STUDIES IN DIFFERENTIATION

The cultured heart preparation has been used to define the time course of the maturation of responsiveness of the fetal mouse heart to a variety of cardioactive drugs and hormones (Wildenthal, 1973a,b; Wildenthal et al., 1976a). For these experiments, hearts from 12- to 22-day-old fetuses were cultured under identical conditions, and the effects of the drug on heart rate and force of contraction were determined. Acetylcholine, for example, caused bradycardia even in 12- to 14-day-old fetal hearts (i.e., prior to cardiac innervation), but glucagon-induced tachycardia was evident only in hearts from fetuses 17 days old or older. This relatively simple experimental protocol identified the time course of the appearance of specific receptors and showed that specific receptors differentiated at different times in the developing heart. Chen et al. (1979) extended these studies by using ligand-binding techniques to quantitate the change in the number of β-adrenergic receptors during development of the mouse heart.

Cultured hearts have also been used to identify factors that might play a role in muscle differentiation. The experiments described above showing that myosin synthesis and CPK activity were greater in creatine-supplied hearts were designed to test the possibility that creatine, which accumulates in hearts during this stage of development, may play a role in stimulating muscle-specific protein synthesis. The results indicate that this may be the case.

2. TRANSPORT SYSTEMS

Cultured hearts have also been used to identify a carrier-mediated transport system for taurine (Grosso *et al.*, 1977, 1978). Uptake of taurine was characterized with respect to temperature dependence, saturability, and structural selectivity. It was found that taurine accumualted against a concentration gradient. The transport system is mediated by the β-amino acid uptake system and is both sodium- and energy-dependent. In order to test such a wide range of experimental conditions, a large number of hearts of the same size and gestational age were required. By using cultured mouse hearts, many replicate hearts could be easily prepared. The protocol used for these studies should have wide application in the study of transprt systems in the heart.

3. THE CULTURED HEART AS A MODEL OF INJURED MYOCARDIUM

An experimental model of total or partial ischemia that exhibits uniformally reversible or irreversible cell injury and in which flow can be controlled would be useful in identifying the characteristics of reversible and irreversible injury. The cultured mouse heart preparation satisfies these requirements at least in part. By manipulating (1) the availability of oxygen and oxidizable substrates, (2) the pH of the culture medium, and (3) the concentration of metabolites known to occur in ischemic tissue, it is possible to impose on these hearts an environment resembling that which exists in the totally ischemic heart *in vivo*. In this context, total ischemia is defined as complete absence of flow, which results in deprivation of oxygen and oxidizable substrates and accumulation of the end products of anerobic metabolism.

In cultured mouse hearts prepared from 17-day-old fetuses and transiently deprived of oxygen and oxidizable substrate (glucose), the time to asystole was short (8 minutes) (Roeske *et al.*, 1976) and preceded measurable changes in ATP concentration (Ingwall *et al.*, 1978). Deprivation for 1 hour did not alter the ATP content of the hearts. When the hearts were resupplied with oxygen and glucose, beating resumed within minutes, and no enzyme depletion (measured as LDH activity) was observed during the following 24 hours of recovery. Deprivation for 2 or 3 hours resulted in depressed ATP levels and some loss of radiolabeled purine nucleosides and bases. When these hearts were resupplied with oxygen and glucose, beating resumed, ATP levels returned to 80–90% of control values, and little or no LDH was released from the heart into the culture medium during the recovery period. Ultrastructural derangements such as disruption of cristae, swelling of mitochondria, and margination of nuclear chromatin occurred early during the period of deprivation but were no longer evident during reoxygenation. With these mechanical, biochemical, and ultrastructural markers, it may be concluded that deprivation for up to 3 hours resulted in reversible injury.

In contrast, deprivation of oxygen and glucose for periods longer than 3 hours resulted in substantial irreversible injury (Ingwall *et al.*, 1975). Such long-term deprivation resulted in 75% reduction in total purine nucleotide content and 40% loss of purine nucleosides and bases from hearts into the culture medium. Within 4–6 hours after resupply of oxygen and glucose, beating was often observed, ATP levels returned to as much as 60% of control values, and many mitochondria appeared to have normal ultrastructure. Resynthesis of ATP occurred solely from preformed bases, nucleosides, and nucleotides (i.e., via the salvage pathway) and not by *de novo* synthesis (Kaufman *et al.*, 1977). During the period of recovery, but not during the period of deprivation, the cytoplasmic enzyme content was depleted by as much as 40% (Roeske *et al.*, 1977, 1978). Thus, recovery from 4–5 hours of deprivation was not complete. When the period of deprivation or insult was extended beyond 5 hours, no recovery was observed.

These observations show that altering the duration of insult in this preparation results in global damage of varying degrees, which resembles that seen in different regions of the focally ischemic heart *in vivo*, ranging from marginally deprived areas capable of full recovery and normal oxidative metabolism to areas undergoing necrosis in the center of the most severe ischemic zone. The ability to control the degree of injury has proved to be useful in providing an understanding of several aspects of myocardial injury.

For example, the uptake of myocardial imaging agents is thought to depend both on the severity of cell injury and the degree of the remaining blood flow. A heart preparation in which the extent of injury can be controlled and which is independent of flow would be useful for defining the various factors determining uptake. Flow-independent uptake of thallous ion and of several technetium-labeled compounds into control, reversibly injured, and irreversibly injured cultured hearts was examined. It was found that thallium uptake occurred only in normal and reversibly injured cells, while technetium-labeled pyrophosphate, glucoheptonate, and tetracycline were taken up primarily by irreversibly injured tissue (Schelbert *et al.*, 1976).

This preparation is particularly well suited for studies on repair mechanisms following transient insult. Changes in several specific metabolic pathways have already been defined, and interventions designed to reverse these changes have been proposed. For example, the time course of purine nucleotide degradation and resynthesis has been defined. It has been found that loss of purine nucleotide precursors from hearts into the culture medium, presumably by diffusion, determines the extent of ATP resynthesis in hearts recovering from transient deprivation. It may be possible to prevent the loss of purines from the heart into the culture medium, and thereby to maintain normal purine pool size, simply by supplying hearts with exogenous purines such as inosine and hypoxanthine. Of course, measuring the average concentration of ATP in the whole heart does not address the important problems of ATP compartmentalization and ATP turnover in normal and injured myocardium. Experiments designed to probe these features of nucleotide metabolism remain to be done. Adenine nucleotide metabolism is an example of just one area of cardiac metabolism that can be profitably studied using cultured mouse hearts. Another is the metabolism of other high-energy-phospahte-containing compounds such as creatine phosphate and GTP in the injured heart. Still another is the regulation of myofibrillar, mitochondrial, and lysosomal protein turnover and function.

B. Future Studies

The spontaneously beating fetal mouse heart in organ culture offers the possibility of obtaining biochemical, functional, and structural information in a well-controlled *in vitro* heart preparation. By assessing a composite set of markers, a greater understanding of the processes important in maintaining normal cardiac function and differentiation may be defined. As in all *in vitro* systems, cultured hearts are maintained under relatively unphysiological conditions, and it is essential to exercise caution in extrapolating results obtained from fetal tissue to the adult myocardium. Nonetheless, it seems to us that this preparation is well suited for exploring many problems in myocardial metabolism, especially those requiring long-term stability and the ability to control experimental conditions. Several examples of this have been described in this chapter. It is anticipated that future work using the cultured fetal mouse will continue to focus on studies of high-energy phosphate metabolism, protein turnover, and repair of injured myocardium. Finally, it should be noted that the techniques described here for culturing fetal mouse hearts serve equally well for culturing fetal heart tissue of other species, including human (Hughes and Longmore, 1972; Longmore and Hughes, 1972). Extension of the types of studies presented in this article to human heart requires only the availability of sufficient biological material.

Acknowledgments

We would like to acknowledge our many collaborators in these studies: K. Billmire, R. Bressler, J. W. Covell, D. S. Grosso, M. DeLuca, E. E. Griffin, N. F. Hall, I. A. Kaufman, S. E. Mayer, H. D. Sybers, R. Watson, and H. I. Yamamura. We also acknowledge the excellent assistance of D. Buccigrossi, S. C. Jascinski, M. Kennedy, P. Kolb, M. Kramer, P. C. Morton, L. Nimmo, and J. R. Wakeland.

REFERENCES

Barrett, A. J. (1972). *In* "Lysosomes: A Laboratory Handbook" (J. T. Dingle, ed.), pp. 46-135. North-Holland Publ., Amsterdam.

Bernstein, L. H., and Everse, J. (1975). *In* "Methods in Enzymology" (W. A. Wood, ed.), Vol. 41, pp. 47-52. Academic Press, New York.

Chen, F. M., Yamamura, H. I., and Roeske, W. R. (1979). *Eur. J. Pharm.* **58**, 255-264.

Fell, H. B., Dingle, J. T., Coombs, R. R. A., and Lachmann, P. J. (1968). *Symp. Int. Soc. Cell Biol.* **7**, 49.

Griffin, E. E., and Wildenthal, K. (1978a). *Am. J. Physiol.* **234**, E306-E313.

Griffin, E. E., and Wildenthal, K. (1978b). *J Mol. Cell. Cardiol.* **10**, Suppl. I, 27.

Grosso, D. S., Roeske, W. R., and Bressler, R. (1977). *Proc. West. Pharmacol. Soc.* **20**, 239-243.

Grosso, D. S., Roeske, W. R., and Bressler, R. (1978). *J. Clin. Invest.* **61**, 944-952.

Hall, N., and DeLuca, M. (1976). *Anal. Biochem.* **37**, 742-751.

Hinegardner, R. T. (1971). *Anal. Biochem.* **39**, 197-201.

Hughes, D. M., and Longmore, D. B. (1972). *Nature* **235**, 334-336.

Ingwall, J. S. (1976). *Circ. Res.* **38**, 115-123.

Ingwall, J. S., and Wildenthal, K. (1976). *J. Cell Biol.* **68**, 159-163.

Ingwall, J. S., and Wildenthal, K. (1978). *Recent Adv. in Stud. Card. Struct. Metab.* **12**, 621-633.

Ingwall, J. S. DeLuca, M., Sybers, H. D., and Wildenthal, K. (1975). *Proc. Natl. Acad. Sci. U.S.A.* **72**, 2809-2813.

Ingwall, J. S., Watson, R., and Mayer, S. E. (1978). *In* "Proceedings of the Third US-USSR Symposium on Cardiac Metabolism" (H. E. Morgan, ed.), pp. 281-287. NIH Publications, Bethesda, Maryland.

Kammermeier, H. (1973). *Anal. Biochem.* **56**, 341-345.

Kaufman, I. A., Hall, N. F., DeLuca, M., Ingwall, J., and Mayer, S. E. (1977). *Am. J. Physiol.* **233**, 282-288.

Kluner, R. A., and Ingwall, J. S. (1980). *Exp. Mol. Pathol.* (in press).

Libby, P., Goldberg, F., and Ingwall, J. S. (1979). *Am. J. Physiol.* **237**, E35-E39.

Longmore, D. B., and Hughes, D. M. (1972). *Nature* **238**, 40-41.

Lowry, O. H., and Passonneau, J. V. (1972). "A Flexible System of Enzymatic Analysis," pp. 151-153. Academic Press, New York.

Lowry, O. H., Rosebrough, N. J., Farr, A. L., and Randall, R. J. (1951). *J. Biol. Chem.* **193**, 265-275.

Maurer, M., Yuhas, D. E., and Miller, J. G. (1979). *J. Mol. Cell. Cardiol.* **11**, 319-323.

Roeske, W. R., Ingwall, J., Kramer, M., and Covell, J. (1976). *Clin. Res.* **24**, 88A.

Roeske, W. R., Ingwall, J. S., DeLuca, M., and Sybers, H. D. (1977). *Am. J. Physiol.* **232**, H288-296.

Roeske, W. R., DeLuca, M., and Ingwall, J. S. (1978). *J. Mol. Cell. Cardiol.* **10**, 907-919.

Rosalki, S. B. (1967). *J. Lab. Clin. Med.* **69**, 696-705.

Schelbert, H., Ingwall, J. S., Sybers, H. D., and Ashburn, W. L. (1977). *Circ. Res.* **39**, 860-868.

Szasz, G., Busch, E. W., and Faroks, H. B. (1970). *Dtsch. Med. Wochenschr.* **95**, 829-835.

Théroux, P., Franklin, D., Ross, J., Jr., and Kemper, W. S. (1974). *Circ. Res.* **35**, 896-908.

Wildenthal, K. (1971). *J. Appl. Physiol.* **20**, 153-157.

Wildenthal, K. (1972). *Nature (London)* **239**, 101-102.

Wildenthal, K. (1973a). *J. Mol. Cell. Cardiol.* **5**, 87-99.

Wildenthal, K. (1973b). *J. Clin. Invest.* **52**, 2250-2258.

Wildenthal, K., and Crie, J. S. (1980). *Fed. Proc.* **39**, 37-41.

Wildenthal, K., and Griffin, E. E. (1976). *Biochim. Biophys. Acta* **444**, 519-524.

Wildenthal, K., Harrison, D. R., Templeton, G. H., and Reardon, W. C. (1973). *Cardiovasc. Res.* **7**, 139–144.

Wildenthal, K., Poole, A. R., and Dingle, J. T. (1975). *J. Mol. Cell. Cardiol.* **7**, 841–855.

Wildenthal, K., Allen, D. O., Karlsson, J., Wakeland, J. R., and Clark, C. M., Jr. (1976a). *J. Clin. Invest.* **57**, 551–558.

Wildenthal, K., Griffin, E. E., and Ingwall, J. S. (1976b). *Circl Res.* **38**, 138–144.

Wildenthal, K., Wakeland, J. R., Morton, P. C., and Griffin, E. E. (1978). *Circl Res.* **42**, 787–792.

Wildenthal, K., Sanford, C. F., Griffin, E. E., and Crie, J. S. (1980). *In* "Advances in Myocardiology" (G. Rona and N. Dhalla, eds.).Univ. Park Press, Baltimore, Maryland (in press).

METHODS IN CELL BIOLOGY, VOLUME 21A

Chapter 12

The Cultured Heart Cell: Problems and Prospects[1]

MELVYN LIEBERMAN, WILLIAM J. ADAM, AND PHYLLIS N. BULLOCK

Department of Physiology,
Duke University Medical Center,
Durham, North Carolina

I. Introduction

The first documented account of heart muscle in tissue culture described the outgrowth of spontaneously beating cells from an explant of embryonic chick heart fragments cultivated in a mixture of blood plasma and embryo extract (Burrows, 1912). There were major problems in using this technique for physiological and biochemical studies. Gaining accessibility to the desired cell type and eliminating necrotic tissue were accomplished by the use of proteolytic enzymes such as trypsin (Moscona, 1952) to dissociate embryonic heart muscle into discrete spontaneously active single cells (Cavanaugh, 1955). The technique of enzymatic dissociation was later applied to the postnatal mammalian heart (Harary and Farley, 1963). Over the last 16 years, this method has been successfully adapted to study the morphology, biochemistry, and physiology of em-

[1]Supported in part by NIH grants HL12157, HL07101, and HL23138.

bryonic, fetal, and neonatal heart muscle, either as dispersed cells or as mono- and multilayered preparations (for reviews, see Lieberman and Sano, 1976; Kobayashi *et al.*, 1978). In recent years, it has been possible to culture enzymatically dispersed embryonic heart cells in tissue-like orientation. These preparations, when subject to a variety of analytical measurements, clearly support the hypothesis that tissue culture methodologies do not alter the physiological properties of heart cells (Lieberman *et al.*, 1972; 1975; Sachs and DeHaan, 1973; Sachs, 1976; McLean and Sperelakis, 1976; Horres *et al.*, 1977).

Over the last 10 years, there have been few studies utilizing human heart cells in culture. Furthermore, this work has not been as successful as that conducted with cardiac muscle cells from nonhuman sources. The results of these studies will be considered in the following section, together with recent results obtained from fetal human heart muscle in our laboratory. In the course of the latter studies, several practical problems emerged that limited the successful implementation of human cardiac cell cultures. The logistics involved in obtaining tissue in good physiological condition and the limited availability of sufficient quantities of heart cells were deterrents in these studies. As a result of these shortcomings, optimal methods have yet to be determined for the disaggregation procedure and the nutritional requirements of human heart cells.

II. Methods and Results

An early attempt to make quantitative measurements on spontaneously beating muscle cells from surgical specimens of adult human myocardium was unsuccessful because too few cells were obtained (Bloom, 1970a). However, cardiac tissue from 7- to 10-week aborted human embryos could be cultured as explants in plasma clots and maintained in a physiologically active state for up to 11 weeks (Chang and Cumming, 1972). In these preparations, sympathomimetic drugs (epinephrine, norepinephrine, isoprenaline), which increased the rate of spontaneous activity, were ineffective in the presence of the β-receptor blocker propranolol. This finding suggested the presence of functional β receptors in human heart cells prior to the third month of development.

The feasibility of cultivating dissociated fetal human heart cells was first established by Halbert *et al.* (1973). Tissue specimens from 13- and 20-week-old fetuses were obtained approximately 30 minutes after surgery. The hearts were minced in calcium- and magnesium-free Hank's solution and then digested in trypsin (0.03–0.04%) with continual agitation for 8–14 incubation periods of 20 minutes' duration. Cells harvested from the last four digestions were then suspended in Eagle's minimal essential medium (MEM) supplemented with 10% fetal calf serum and cultured in Rose chambers at densities of 2–3 × 10^6 cells/ml.

Within 24 hours, approximately half of the inoculum attached to the glass sub-strate and 20% of the cells were observed to beat spontaneously at a rate of 30 beats/minute. Synchronously beating monolayers (60 beats/minute) were ob-served after 3 days in culture, and striations were evident in a number of the contracting cells. In several instances, mitotic activity was noted in beating myocardial cells. Frequent changes of the culture medium enabled these inves-tigators to maintain some of the cultures for 3 weeks. Nevertheless, cytoplasmic vacuoles appeared in growing numbers unless the medium was supplemented with whole human serum. This study was essentially repeated by Thompson (1977), using cells obtained from a 20-week human fetal heart to demonstrate the susceptibility of heart cells (muscle and nonmuscle) to diphtheria toxin.

In an attempt to evaluate the problems associated with establishing primary cultures of human heart cells, we applied and modified well-established methods for the culture of embryonic avian heart cells in our laboratory (cf. Lieberman *et al.*, 1975). On two occasions, fetal hearts, approximately 14–16 weeks of age, were obtained within 2 hours after surgery and transported to the laboratory in cold medium 199 (M199). The tissue was minced, suspended in calcium-free, magnesium-free Hank's balanced salt solution containing 0.05–0.1% trypsin (1–300, NBC), subjected to three or four 8-minute disaggregation cycles at 37°C. During this time the tissue mince was continuously agitated. The cells were resuspended in M199 (containing modified Earle's salt solution) supplemented with 2% fetal calf serum (heat-inactivated) and 2% chick embryo extract. Since sterile techniques were used throughout the procedure and all work was carried out in a laminar-flow hood, the culture medium was not supplemented with antibiotics. Culture dishes (60 mm diameter, Falcon 3001) were seeded with $1.5–2.0 \times 10^6$ cells/ml and incubated at 37°C for up to 5 days in a humidified atmosphere of 4% carbon dioxide in air. After 24 hours, cell attachment was generally poor, and few cells were spontaneously beating. The situation im-proved somewhat after 3 days in culture, although the results were markedly different from those routinely observed with embryonic avian heart cells. Phase-contrast micrographs from cultures after 3 and 4 days' incubation (Fig. 1) illus-trate the presence of aggregates of heart cells (refractile clusters) which were spontaneously contracting at rates of 40 beats/minute. In general, contracting cells were associated with beating aggregates, and the question as to whether these cells were pacemaking or stimulated by the aggregates is unresolved.

After 5 days in culture, preparations were fixed for electron microscopic observation. Unless otherwise noted in the figure legends, preparation of mate-rials for electron microscopy proceeded as follows. The tissue was rinsed four times in M199, followed by fixation in 0.83% sodium phosphate-buffered glutaraldehyde with an osmolarity adjusted to 280 mosm/kg. After three changes (2 minutes each) of M199, the tissue was postfixed in 1% sodium cacodylate buffer. After one rinse with M199 (2 minutes) the specimens were

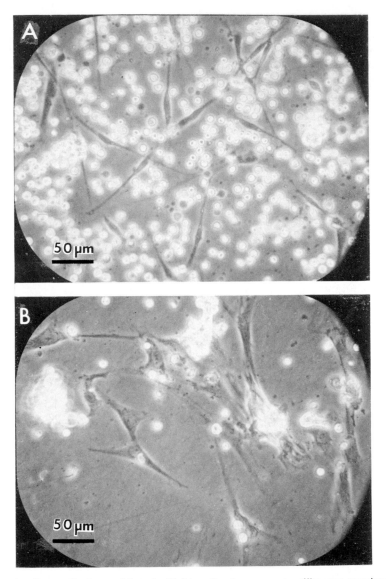

FIG. 1. Heart cells from a 14-week-old fetus. Spontaneous contractility was associated with either floating or attached aggregates (refractile in appearance). (A) Three days in culture. (B) Four days in culture. ×200.

dehydrated in a graded series of ethanol and infiltrated with Epon 812. All procedures were carried out at 37°C. Postosmification and dehydration were carried out at 0°C. Infiltration was performed in a dessicator under a low vacuum. Figure 2 shows sections from a beating cluster of cells and illustrates relatively immature sarcomere organization with remnants of sarcoplasmic reticulum, a transverse tubular system, and an intact basal lamina.

Studies to date with adult and fetal human heart cells in tissue culture are at best preliminary, since several critical problems remain to be solved. From the standpoint of logistics, the availability of tissues in quantities sufficient to establish good cultures is limited. Furthermore, establishing a schedule for obtaining enough tissue in satisfactory physiological condition is not trivial, since it is difficult to predict the time at which specimens of tissue will become available. Last, the results obtained illustrate the inadequacy of the methods presently used to dissociate and maintain human heart cells in culture. These problems certainly are neither formidable nor insurmountable and should be resolved as the experimental needs for human heart cells become more clearly focused and defined.

In the context of this presentation, it is appropriate to call attention to several studies that have reported results obtained from a cell line of human heart cells derived from a biopsy specimen of right atrial appendage (Girardi et al., 1958). This cell line has been used in several studies directed toward gaining information about certain transport properties of the cell membrane, particularly sodium and potassium exchange (Lamb and McCall, 1972; McCall, 1976a,b) and carnitine uptake (Bohmer et al., 1977; Molstad et al., 1978). Although Girardi heart cells have been shown to be a stable preparation with membrane transport properties similar to those of other mammalian tissues, the reported results should not be related to those obtained from intact or cultured preparations of cardiac muscle. Contrary to electrically excitable cardiac cells, Girardi heart cells are electrically inexcitable and have a low resting membrane potential of -20 mV (Lamb and McCall, 1972). Recent ultrastructural studies have established that Girardi heart cells lack myofilaments and have a cytoplasmic appearance characteristic of nonmuscle cells (Molstad et al., 1978). Hence this cell line is neither functionally nor morphologically comparable to the cell type of origin. Since the Girardi heart cell line has also been found to contain HeLa marker chromosomes (Lavappa, 1978) and the type-A isoenzyme of glucose-6-phosphate dehydrogenase (R. J. Hay, personal communication), the transport properties of the cell membrane must be considered to be more like those of inexcitable (nonmuscle) cells in tissue culture.

III.   Discussion and Perspectives

The problems associated with the cultivation of freshly excised human heart cells could perhaps be alleviated if a myocardial cell line of functionally active cells were available. Since differentiated heart cells from several species (rat,

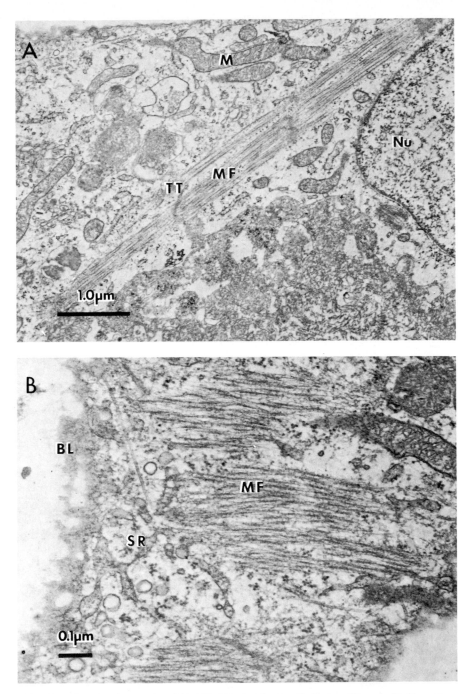

FIG. 2. Electron micrographs of identified beating heart cells from a 16-week-old fetus after 5 days in culture. Note the relatively immature myofibrillar organization and remnants of the myofilaments (MF), transverse tubules (TT), and sarcoplasmic reticulum (SR). M, Mitochondria; BL, basal lamina; Nu, nucleus. (A) ×20,000. (B) ×90,000.

chick, human) have been shown to undergo mitosis (Kasten, 1972; Chacko, 1973; Halbert *et al.*, 1973), the establishment of a line of beating heart cells could provide investigators with an ideal preparation because it would ensure a continuous supply of homologous cells. In 1970 Paul reported that it was possible to subculture beating heart cells from a 2-day-old rabbit embryo (see Wollenberger, 1978), but the details of this study do not appear in the literature. A continuous cell line from the heart of the box turtle was successfully cultivated but, by the sixth passage, the cells were entirely epithelial-like (Clark and Karzon, 1967). It is of interest to note that Kimes and Brandt (1976) derived a clonal cell line from embryonic BDIX rat heart muscle that exhibited morphological and biochemical properties of skeletal muscle. More recently, a cell line isolated from mouse teratocarcinoma (Amano *et al.*, 1978) was induced to differentiate into rhythmically contracting heart muscle cells.

A.   Current Applications

The prospect of establishing primary cultures of adult human heart cells is beginning to emerge as a distinct possibility, because methods have been established for isolating other mammalian adult heart cells. Using a combination of enzymes that included trypsin, α-chymotrypsin, and collagenase, Kono (1969) obtained a well-dispersed cell suspension of adult rat heart cells. Vahouny *et al.* (1970) prepared a suspension of cardiac cells from adult rat hearts by prolonged incubation with trypsin and collagenase. Morphologically intact muscle cells were obtained by perfusing adult rat hearts with a salt solution containing 0.1% collagenase and 0.2% hyaluronidase (Berry *et al.*, 1970). Although these early studies demonstrated the feasibility of isolating beating cells from adult rat hearts, the preparations were characteristically short-lived (few hours) and demonstrated a low tolerance to calcium, i.e., underwent a maintained contraction in its presence. However, recent techniques have clearly shown that cells can be isolated from the adult rat heart in a manner that promotes the retention of several biochemical (Moustafa *et al.*, 1976; Powell and Twist, 1976; Cutilletta *et al.*, 1977; Farmer *et al.*, 1977; Grosso *et al.*, 1977; Rajs *et al.*, 1978; Carlson *et al.*, 1978), morphological (Nag *et al.*, 1977; Powell *et al.*, 1978c; Carlson *et al.*, 1978; Rajs *et al.*, 1978; Moses and Kasten, 1979), and physiological (Fabiato and Fabiato, 1973; Grosso *et al.*, 1977; Carlson *et al.*, 1978) characteristics. Specifically, isolated heart cell preparations can be maintained in a viable state for 24 hours. During this time, they actively synthesize RNA and oxidize carbohydrates, lipids, and fatty acids, show respiratory coupling and a tolerance for calcium chloride, and have an intact sarcolemma responsive to β-adrenergic stimulation, insulin, and glucagon. Of notable significance is the fact that physiologists have recently succeeded in recording transmembrane action potentials from these preparations (Powell *et al.*, 1978a,b) and, in one instance, have utilized a technique that allows voltage clamp and internal perfusion (Lee *et al.*, 1979). Adult cardiac myocytes have also been isolated from the hearts of mouse (Bloom, 1970b; Tsokos and Bloom, 1977), dog (Lieu and Spitzer, 1978; Vahouny *et al.*, 1979), rabbit (Dani *et al.*, 1977), squirrel (Lyman and Jarrow, 1971), and frog (Tarr and Trank, 1976). Future directions have recently been delineated by the report of a method for isolating single myocardial cells from adult rats and establishing them in culture for up to 60 days (Jacobson, 1977). In a subsequent report (Kennedy and Jacobson, 1979), membrane potentials ranging in amplitude from 40 to 70 mV were recorded from spontaneously active cells (Fig. 3).

The application of methods for isolating and culturing adult mammalian heart cells should be extended to human cardiac muscle in the coming decade. The possibility of obtaining large numbers of cells from limited supplies of tissue has been further enhanced by the availability of methods for preserving isolated adult heart cells at $-196°C$ (Alink *et al.*, 1977) and satisfactorily growing them in culture after subsequent thawing (Alink *et al.*, 1978). However, the missing link in the maintenance of stable differentiated cultures of human heart cells probably is a better understanding of their nutritional requirements. In this regard, two amino acids present in very high concentrations in heart muscle, taurine (Scharff and Wool, 1965) and carnitine (Marquis and Fritz, 1965), may prove to be necessary components for preparing a satisfactory defined culture medium for human heart muscle.

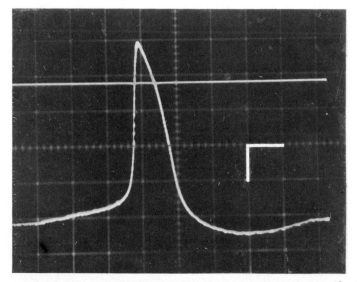

Fig. 3. Transmembrane action potential recorded from adult rat heart cells in culture for 16 days. Horizontal line indicates zero potential. Calibration: vertical bar, 10 mV; horizontal bar, 50 msec. (Courtesy of Dr. S. Jacobson, Ottawa.)

B.   Future Studies

In this section we will highlight examples of the kind of preparations that could be developed with human heart cells once the methods for successfully establishing them in culture are resolved. Although these preparations were developed specifically for studying the physiological properties of cardiac muscle, it should not be difficult to envision their applicability to anatomical, biochemical, and pathological studies.

Unable to find a naturally occurring preparation of cardiac muscle that satisfied our criteria of simplicity (Johnson and Lieberman, 1971), we explored the feasibility of growing, in tissue culture, a preparation of cardiac muscle that would conform to a desired geometry (cylindrical cable or sphere). We developed a differentiated preparation of cells from embryonic chick hearts, cultivated in such a way that tiny bundles of fibers could be grown to any desired length (Lieberman *et al.*, 1972). The "synthetic strand" represents a reconstruction of cardiac muscle fibers with a linear, relatively simple geometry (Fig. 4). Electron micrographs of the strands clearly demonstrated that, by a number of ultrastructural criteria, the muscle cells differentiated normally (Purdy *et al.*, 1972; Lieberman *et al.*, 1973b). Analysis of the linear electrical properties of synthetic strands (e.g., Fig. 4A) showed them to behave as one-dimensional,

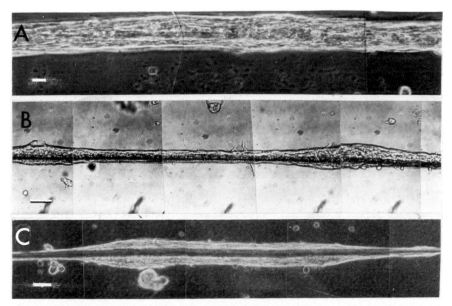

FIG. 4.   Photomicrographs of embryonic chick heart cells grown in tissue culture along a channel cut in the agar substrate. (A) Segment of a long synthetic strand. ×80. (B) Segment of a synthetic strand exhibiting nonuniform conduction between the two bulbous regions. ×120. (C) Short synthetic strand. ×100.

leaky capacitive cables whose specific electrical constants were R_m, 20 Kohm·cm²; C_m, 1.5 μF/cm²; R_i, 180 ohm·cm (Lieberman *et al.*, 1975). Although preparations with properties of nonuniform conduction were generally associated with progressive dimensional changes along their length (Fig. 4B), the delays in propagation were attributable to alterations in coupling resistance (Lieberman *et al.*, 1973a,b) and inactivation of the sodium-carrying mechanism (Lieberman *et al.*, 1976). Cable complications in the analysis of membrane properties by the conventional two-electrode voltage clamp technique demanded a refinement of our tissue culture techniques in order to produce short preparations of reduced diameter. By reducing the length of the growth-orienting channels to approximately 1.0 mm and limiting the number of rapidly dividing nonmuscle cells in the original cell suspension, we were able to obtain short (approximately 200 μm), relatively narrow (50–100 μm) strands that were ovoid in shape (Fig. 4C) and suitable for voltage clamp analysis (Lieberman *et al.*, 1976).

The short synthetic strand appeared to behave as a single, one-dimensional cell

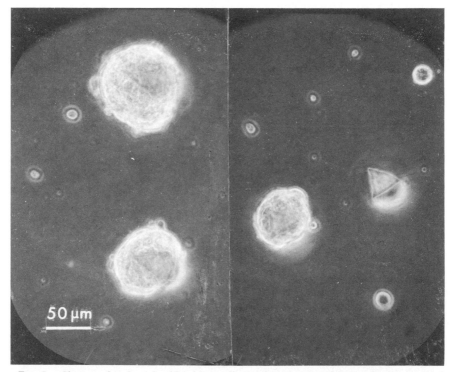

Fɪɢ. 5.   Clusters of embryonic chick heart cells attached to exposed areas of an agar-coated culture dish. ×240.

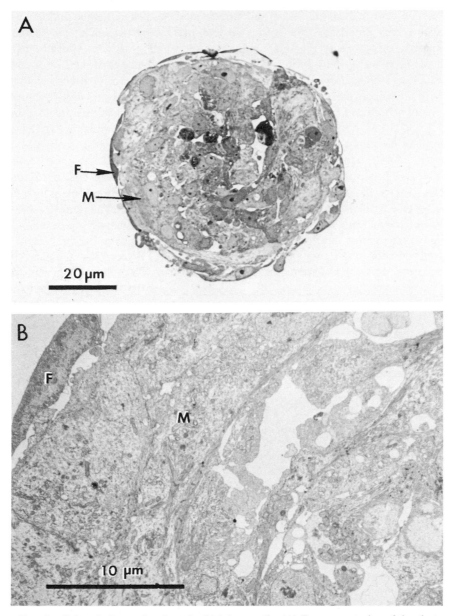

FIG. 6.   Cultured cluster of embryonic chick heart cells. (A) Transverse section of the cluster depicting the compact arrangement of the inner core of muscle cells (M) and outer sheath of fibroblast-like cells (F). ×900. (B) Electron micrograph of the cultured cluster demonstrates the normal configuration of the muscle (M) and fibroblast-like (F) cells. The myocytes appear to be more tightly packed toward the periphery of the preparation. ×3600.

for slowly changing currents, and subsequent studies showed that the multicellularity of the preparation per se did not contribute significantly to the spatial inhomogeneity of the potential within the preparation (R. A. Chilson, M. Lieberman, and E. A. Johnson, unpublished observations). However, we found that short strands grown in grooves could link to one another by processes that were often invisible, even at high magnification, because of the superimposed effects of light diffraction caused by the groove itself. It was then decided to develop a preparation that came closest to the ideal in minimizing problems associated with using microelectrodes as point sources of current (Eisenberg and Engel, 1970), namely, a spherical aggregate or cultured cluster of heart cells (Fig. 5). Small (25-μm) openings were chipped out of an agar-coated culture dish using a 27-gauge needle. The exposed plastic surface was sufficient in area to provide a point of attachment for preformed small, spherical aggregates (50–200 μm diameter). Only preparations less than 100 μm in diameter were selected for voltage clamp analysis using two microelectrodes (Ebihara *et al.*, 1980). Figure 6 shows light and electron micrographs of a cultured cluster typical of those used for such analyses. The preparation consists of a central core of muscle cells enveloped by a single cell layer of flattened fibroblast-like cells. To bypass

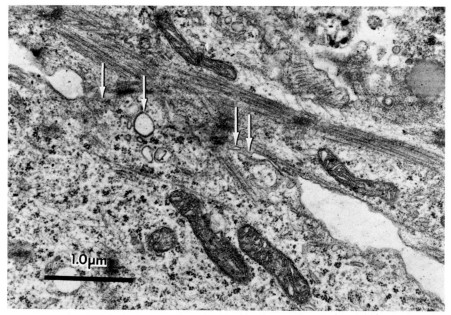

FIG. 7. Electron micrograph of a liposome-treated preparation of embryonic chick heart cells. Cell fusion is suggested by the presence of membrane discontinuities and cytoplasmic bridging (arrows). (Processed for electron microscopy according to Horres *et al.*, 1977.) ×23,000.

limitations of analysis that may result from studies using a multicellular preparation (ion accumulation and/or depletion in extracellular space), our recent efforts have been directed toward inducing fusion of cultured cardiac muscle cells to produce single, multinucleate, spherical cells. Because of problems associated with the use of lysolecithin (Lieberman *et al.*, 1976), heart cells were treated with liposomal vesicles prepared according to the method of Martin and McDonald (1976). Small, sonicated, unilamellar vesicles comprised of egg lecithin, lysolecithin, and stearylamine (prepared by S. Simon) were added to a suspension of cells in a balanced salt solution. Multinucleate arrays of beating heart cells were identified and fixed for electron microscopic examination. Figure 7 shows evidence of membrane discontinuities and cytoplasmic bridging suggestive of membrane fusion. Since the incidence of fusion with this method is somewhat limited, we are currently pursuing alternative methods for promoting the fusion of cardiac cells.

Identification of the different ion species associated with the electrical activity of cardiac muscle is concomitantly being pursued by making isotopic flux measurements of another type of strand preparation. Since a relatively weak adhesive force exists between the synthetic strands (mentioned above) and the culture

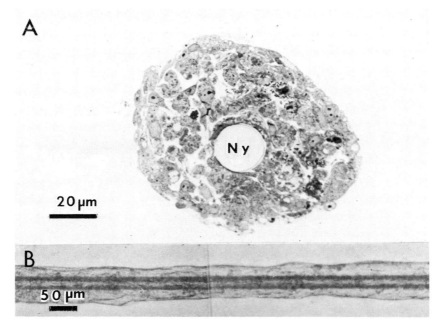

FIG. 8. Growth-oriented heart cells around a nylon monofilament. (A) Photomicrograph of a cross section of a contractile preparation. Ny, Nylon monofilament. ×600. (B) Phase-contrast photomicrograph of a segment of a contractile preparation. ×150.

surface, the use of rapidly flowing solutions is precluded for such studies. We succeeded, however, in developing a multistranded preparation of cultured heart cells grown as thin annuli around a nylon monofilament (Horres *et al.,* 1977). This provided a mechanically stable, electrically stimulatable preparation. Figure 8 illustrates the short diffusional distance and morphologically simple extracellular space of the preparation. The ability to increase tissue mass without comprising surface area proved to be of unquestionable value for quantitative determinations of ^{42}K transport across the cell membrane (Horres and Lieberman, 1977; Horres *et al.,* 1979). A major question arising from the radioisotope flux and ion content studies of heart cells concerned the extent to which a decrease in the activity of the sodium–potassium transport system (Na-K pump) is implicated in the mechanism(s) responsible for the ultrastructural derangement and functional incompetence of the cells in ischemic myocardium. Inhibition of active transport by perfusing preparations with a potassium-free solution for 15 minutes or by the addition of 10^{-4} *M* ouabain for 30 minutes was sufficient to alter the homeostatic electrochemistry of the cells (Horres *et al.,* 1979) and to induce the accumulation

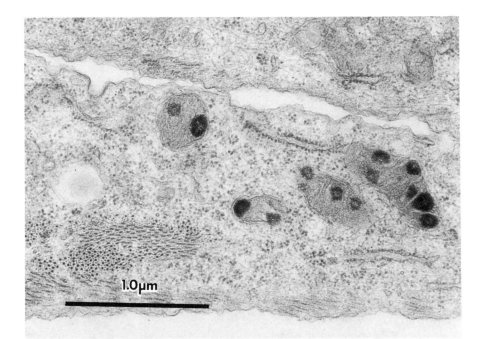

Fig. 9.   Electron micrograph of a preparation of cultured heart cells treated with ouabain (10^{-4} *M*) for 30 minutes. Electron-dense deposits are evident in the mitochondria. ×38,000. (Processed for electron microscopy according to Horres *et al.,* 1977.)

of electron-dense material in the mitochondria (Fig. 9). The significance of this finding, suggesting calcium accumulation in the mitochondria, and its relation to the cellular mechanisms associated with the onset of myocardial ischemia remain to be determined.

In conclusion, studies performed with growth-oriented embryonic heart cells in tissue culture have provided us with fundamental information about membrane permeability, active transport, and correlative morphological and biochemical information. When feasible, the application of such methods to human heart tissue in culture should extend our current understanding of the fundamental processes associated with the normal and pathophysiological behavior of adult cardiac muscle.

ACKNOWLEDGMENTS

We thank Dr. Selman Welt for providing us with fetal heart specimens, and Ms. Pam Evans for technical assistance.

REFERENCES

Alink, G. M., Verheul, C. C., and Offerijns, F. G. J. (1977). *Cryobiology* **14**, 339–448.
Alink, G. M., Verheul, C. C., Agterberg, J., Feltkamp-vroom, T. M., and Offerijns, F. G. J. (1978). *Cryobiology* **15**, 44–58.
Amano, S., Uno, K., and Hagiwara, A. (1978). *Dev., Growth & Differ.* **20**, 41–47.
Berry, M. N., Friend, D. S., and Scheur, J. (1970). *Circ. Res.* **26**, 679–687.
Bloom, S. (1970a). *Comp. Biochem. Physiol.* **37**, 127–129.
Bloom, S. (1970b). *Science* **167**, 1727–1729.
Bohmer, T., Eiklid, K., and Jonsen, J. (1977). *Biochim. Biophys. Acta* **465**, 627–633.
Burrows, M. T. (1912). *Muench. Med. Wochenschr.* **59**, 1473–1475.
Carlson, E. C., Grosso, D. S., Romero, S. A., Frangakis, C. J., Byus, C. V., and Bressler, R. (1978). *J. Mol. Cell. Cardiol.* **10**, 449–459.
Cavanaugh, M. W. (1955). *J. Exp. Zool.* **128**, 573–589.
Chacko, S. (1973). *Dev. Biol.* **35**, 1–18.
Chang, T. D., and Cumming, G. R. (1972). *Circ. Res.* **30**, 628–633.
Clark, H. F., and Karzon, D. T. (1967). *Exp. Cell Res.* **48**, 263–268.
Cutilletta, A. F., Aumont, M.-C., Nag, A. C., and Zak, R. (1977). *J. Mol. Cell. Cardiol.* **9**, 399–407.
Dani, A. M., Cittadini, A., Flamini, G., Festuccia, G., and Terranova, T. (1977). *J. Mol. Cell. Cardiol.* **9**, 777–784.
Ebihara, L., Shigeto, N., Lieberman, M., and Johnson, E. A. (1980). *J. Gen. Physiol.* **75** (in press).
Eisenberg, R. S., and Engel, E. (1970). *J. Gen. Physiol.* **55**, 736–757.
Fabiato, A., and Fabiato, F. (1973). *Eur. J. Cardiol.* **1**, 143–155.
Farmer, B. B., Harris, R. A., Jolly, W. W., Hathaway, D. R., Katzberg, A., Watanabe, A. M., Whitlow, A. L., and Besch, H. R., Jr. (1977). *Arch. Biochem. Biophys.* **179**, 545–558.

Girardi, A. J., Warren, J., Goldman, C., and Jeffries, B. (1958). *Proc. Soc. Exp. Biol. Med.* **98**, 18–22.

Grosso, D. S., Frangakis, C. J., Carlson, E. C., and Bressler, R. (1977). *Prep. Biochem.* **7**, 383–401.

Halbert, S. P., Bruderer, R., and Thompson, A. (1973). *Life Sci.* **13**, 969–975.

Harary, I., and Farley, B. (1963). *Exp. Cell Res.* **29**, 466–474.

Horres, C. R., and Lieberman, M. (1977). *J. Membr. Biol.* **34**, 331–350.

Horres, C. R., Lieberman, M., and Purdy, J. E. (1977). *J. Membr. Biol.* **34**, 313–329.

Horres, C. R., Aiton, J. F., and Lieberman, M. (1979). *Am. J. Physiol.* **236**, C163–C170.

Jacobson, S. L. (1977). *Cell Struct. Funct.* **2**, 1–9.

Johnson, E. A., and Lieberman, M. (1971). *Annu. Rev. Physiol.* **33**, 479–532.

Kasten, F. H. (1972). *In Vitro* **8**, 128–149.

Kennedy, C. B., and Jacobson, S. L. (1979). *J. Mol. Cell. Cardiol.* **11**, Suppl. 1, 31 (abstr.).

Kimes, B. W., and Brandt, B. L. (1976). *Exp. Cell Res.* **98**, 367–381.

Kobayashi, T., Ito, Y., and Rona G., eds. (1978). ''Recent Advances in Studies on Cardiac Structure and Metabolism,'' Vol. 12. Univ. Park Press, Baltimore, Maryland.

Kono, T. (1969). *Biochim. Biophys. Acta* **178**, 397–400.

Lamb, J. F., and McCall, D. (1972). *J. Physiol. (London)* **225**, 599–617.

Lavappa, K. S. (1978). *In Vitro* **14**, 469–475.

Lee, K. S., Weeks, T. A., Kao, R. L., Akaike, N., and Brown, A. M. (1979). *Nature (London)* **278**, 269–271.

Lieberman, M., and Sano, T., eds. (1976). ''Developmental and Physiological Correlates of Cardiac Muscle.'' Raven Press, New York.

Lieberman, M., Roggeveen, A. E., Purdy, J. E., and Johnson, E. A. (1972). *Science* **175**, 909–911.

Lieberman, M., and Kootsey, J. M., Johnson, E. A., and Sawanobori, T. (1973a). *Biophys. J.* **13**, 37–55.

Lieberman, M., Manasek, F. J., Sawanobori, T., and Johnson, E. A. (1973b). *Dev. Biol.* **31**, 380–403.

Lieberman, M., Sawanobori, T., Kootsey, J. M., and Johnson, E. A. (1975). *J. Gen. Physiol.* **65**, 527–550.

Lieberman, M., Sawanobori, T., Shigeto, N., and Johnson, E. A. (1976). *In* ''Developmental and Physiological Correlates of Cardiac Muscle'' (M. Lieberman and T. Sano, eds.), pp. 139–153. Raven Press, New York.

Liu, M., and Spitzer, J. J. (1978). *J. Mol. Cell. Cardiol.* **10**, 415–426.

Lyman, C. P., and Jarrow, E. L. (1971). *Exp. Cell Res.* **68**, 214–217.

McCall, D. (1976a). *Cardiovasc. Res.* **10**, 537–548.

McCall, D. (1976b). *J. Pharmacol. Exp. Ther.* **197**, 605–614.

McLean, M. J., and Sperelakis, N. (1976). *Dev. Biol.* **50**, 134–141.

Marquis, N. R., and Fritz, I. B. (1965). *J. Biol. Chem.* **240**, 2193–2200.

Martin, F., and MacDonald, R. (1976). *J. Cell Biol.* **70**, 494–505.

Molstad, P., Bohmer, T., and Hovig, T. (1978). *Biochim. Biophys. Acta* **512**, 557–565.

Moscona, A. (1952). *Exp. Cell Res.* **3**, 535–539.

Moses, R. L., and Kasten, F. H. (1979). *J. Mol. Cell. Cardiol.* **11**, 161–172.

Moustafa, E., Skomedal, T., Osnes, J., and Oye, I. (1976). *Biochim. Biophys. Acta* **421**, 411–415.

Nag, A. C., Fischman, D. A., Aumont, M. C., and Zak, R. (1977). *Tissue & Cell.* **9**, 419–436.

Powell, T., and Twist, V. W. (1976). *Biochem. Biophys. Res. Commun.* **72**, 327–333.

Powell, T., Terrar, D. A., and Twist, V. W. (1978a). *J. Physiol. (London)* **282**, 23P.

Powell, T., Terrar, D. A., and Twist, V. W. (1978b). *J. Physiol. (London)* **284**, 148P.

Powell, T., Steen, E. M., Twist, V. W., and Woolf, N. (1978c). *J. Mol. Cell. Cardiol.* **10**, 287–292.

Purdy, J. E., Lieberman, M., Roggeveen, A. E., and Kirk, R. G. (1972). *J. Cell Biol.* **55**, 563–78.

Rajs, J., Sundberg, M., Sundby, G., Danell, N., Tornling, G., Biberfeld, P., and Jakobsson, S. W. (1978). *Exp. Cell Res.* **115**, 183–189.

Sachs, F. (1976). *J. Membr. Biol.* **28**, 373–399.

Sachs, H. G., and DeHaan, R. L. (1973). *Dev. Biol.* **30**, 233–240.

Scharff, R., and Wool, I. G. (1965). *Biochem. J.* **97**, 257–271.

Tarr, M., and Trank, J. W. (1976). *Experientia* **32**, 338–339.

Thompson, A. (1977). *J. Mol. Cell. Cardiol.* **9**, 945–956.

Tsokos, J., and Bloom, S. (1977). *J. Mol. Cell. Cardiol.* **9**, 823–836.

Vahouny, G. V., Wei, R., Starkweather, R., and Davis, C. (1970). *Science* **167**, 1616–1618.

Vahouny, G. V., Wei, R. W., Tamboli, A., and Albert, E. N. (1979). *J. Mol. Cell. Cardiol.* **11**, 339–357.

Wollenberger A. (1978). *Recent Adv. Stud. Card. Struct.* **12**, 605–608.

Chapter 13

Organ and Cell Culture Applied to the Cardiovascular System—An Overview

EARL P. BENDITT

Department of Pathology,
University of Washington,
Seattle, Washington

Culture of cells *in vitro* from various tissues and organs is now established, and detailed applications to different cell types in various systems are evolving rapidly. The method is powerful, and yields of information on growth requirements and characteristics of certain cell types derived from the artery wall are emerging rapidly (see Chapters 10 and 11). There are certain features that a priori must alter the performance of single cell types in culture. For example, cells in culture are generally forced to undergo frequent and rapid cell multiplication in a fashion not demanded of cells in most organs or tissues, especially the artery walls and heart which have a low mitotic turnover. Changes in morphology and in associated metabolic features and chemical composition with time and passage generation are seen in smooth muscle cells of artery walls in culture (see Chapters 10 and 11). These deviations indicate that understanding of the *in vivo* functioning of cells in the organized tissue structure must be sought after in systems more closely resembling the true life situation. Thus there is a real need to pursue the development of properly controlled organ culture systems with which to assess the function and interactions of the individual cell types in the composite integrated tissue.

It is clear from the foregoing chapters that organ culture is feasible for the vascular system. Rings, segments, and tissue bits from artery walls of rabbits and human beings can be maintained in a viable state (by morphological and metabolic criteria) for periods of a week or more under certain conditions (Chapters 9 and 11). Refinements in conditions and extensions of morphological and biochemical assays with the aim of better approximation to physiological conditions seem within reach. By combining cell culture with organ culture and examining the similarities of these with appropriate *in vivo* studies a better appreciation of how the artery wall operates and what the nature of its alterations are in different disease states should be forthcoming in the forseeable future. It will take a large measure of patient persistence, understanding, and appreciation of the need for these kinds of studies to accomplish the task.

With an organ larger than the artery wall, fed not only by diffusion, but also by its properly functioning circulation, the problem becomes more difficult. However, at least as applied to the developing mouse heart (Chapter 12), the feasibility of *in vitro* cardiac organ culture for studies of functional properties of the embryonic heart is established. Some important metabolic, physiological, and pathological changes induced by manipulation of the culture environment have been made. These give one great hope for future developments in studies of acute and chronic injury to the heart and their prevention or amelioration.

Chapter 14

Explant Methods for Epidermal Cell Culture[1]

SUSAN M. FISCHER, AURORA VIAJE, GERALD D. MILLS, AND THOMAS J. SLAGA

Biology Division,
Oak Ridge National Laboratory,
Oak Ridge, Tennessee

[1]Research supported jointly by the National Cancer Institute contract YO1CP70227 and the Office of Health and Environmental Research, U.S. Department of Energy, under contract W-7405-eng-26 with the Union Carbide Corporation.

I.   Introduction

Culturing human skin is among the oldest of tissue culture techniques. The possibility of keeping excised tissue alive for more than a few days was first explored in 1898 by Ljunggren when he demonstrated by reimplantation that human skin could survive *in vitro* for many days under appropriate conditions (Paul, 1961). The techniques used then are those that are in general use today, although the development of specific media, tissue culture ware, and controlled environments has changed haphazard success to a repeatable process in which skin cultures can be used as a tool in measuring various parameters of cell growth and maturation.

Full-thickness human skin is composed of epidermis, dermis, and subcutaneous tissue. Dermal components such as fibroblasts have been useful in studies on somatic cell genetics, cellular aging, and transformation. Keratinocytes, however, are more pertinent because of the differentiation process they undergo and because the majority of human cancers arise from epithelial cells. The extent of epidermal outgrowth from an explant of skin has been used in experiments to study viral and fungal growth; the effect of allergens (Karasek, 1966), cortisol, and vitamins on epidermal growth (Prose *et al.,* 1967); the toxic effects of antibiotics, topical medication, and chemicals (Karasek, 1966; Prose *et al.,* 1967); and proliferation in psoriasis (Pullman *et al.,* 1974). Cultures of epidermal cells are basically of two types: either explant cultures or outgrowths from small blocks of tissue, or dispersed cell cultures from epidermal–dermal dissociated tissues. The advantages and applications of the explant method are numerous. The techniques and accouterments needed for such cultures are minimal, simple, and well-suited to all sources of skin. This method lends itself to the production of large numbers of cells and/or cultures from relatively small samples of skin. Because there is no requirement for a feeder layer, it is possible to study growth factors, effects of pharmacological agents, and so forth, without having to contend with possible metabolic contributions from a feeder layer. Furthermore, maintenance of normal skin architecture through the use of tissue blocks may aid in initiating an epidermal culture that mimics *in vivo* states of growth and maturation.

The primary goal of this chapter is to discuss explant methodology, culture conditions, and growth requirements and to relate cytological characteristics of the culture to whole skin. The usefulness and drawbacks of this approach to culturing human epidermis will be described for pharmacological studies, carcinogenesis, and genetics.

II.   Materials and Methods

A.   Tissue Source and Collection

Foreskins, the most readily available source of human skin, were obtained weekly from a local hospital. At the time of removal the skins were placed in sterile containers of Eagle's basal medium (BME) supplemented with 10% fetal calf serum (FCS) and high levels of Fungizone; they were kept refrigerated. The data presented in this chapter were derived from newborn foreskin cultures, although comments will be made with regard to adult skin.

Adult skin was obtained during such procedures as cosmetic surgery and grafting for burn patients. Punch biopsies can also be procured. Storage conditions should be the same as for foreskins, although the adult skin sections supplied by the hospital were wrapped in saline-moistened sterile gauze and had been stored refrigerated for 1–4 weeks.

B.   Tissue Preparation

The tissue was soaked in sterile phosphate-buffered saline (PBS) containing Fungizone for several hours prior to preparation for culture. As yeast is the most common contaminant, antibiotics such as penicillin and streptomycin (Penn-Strep) were rarely used at this point. After soaking, the foreskins were cut to lie flat, and the subcutaneous tissue and as much of the dermis as possible were removed with scissors. A scalpel was used to scrape the dermis lightly and trim off the edges. After a rinse in PBS, the tissue was stretched dermal side down on a large plastic petri dish. Small (1.5 × 1.5 mm) blocks were cut from the skin with a scalpel or razor blade. The blade was very sharp so that the tissue would not be mashed. The blocks were transferred to a 60-mm tissue culture dish and placed in a 2 × 2 pattern according to a predrawn arrangement on a piece of paper placed under the dish; this ensured equal spacing between blocks and identical spacing in all dishes. When the blocks or explants lost their wet appearance (but before they were completely dried out), 3 ml of complete medium was added to the dish. Incubation was at 37°C and 5% CO_2 unless otherwise indicated. Medium was routinely changed every 3–4 days. Epidermal outgrowth was monitored microscopically, although growth beyond 1 mm was visible to the unaided eye. Although not common, fibroblast outgrowth occurs under some conditions and can be controlled by either removal with trypsin or removal by scraping and aspiration. In the studies presented here, no effort was made to remove fibroblasts.

C.   Culture Conditions and Treatment

The medium used was either BME with Hank's salts, supplemented with 10% FCS (unless otherwise indicated), 10,000 U penicillin plus 10,000 μg streptomycin (Penn-Strep) per liter, and 1.25 mg Fungizone per liter; or an enriched Waymouth's MB 752/1 [referred to here as Super Medium (SM)] containing 10% FCS, 1× nonessential amino acid solution, 457 mg L-arginine per liter, 220 mg sodium pyruvate per liter, 0.322 mg putrescine–HCl per liter, 10.0 mg bovine insulin per liter, 0.1 mg hydrocortisone per liter, 2240 mg $NaHCO_3$ per liter, and Penn-Strep and Fungizone as above. All media and antibiotics were purchased from Grand Island Biological Company; FCS was obtained from either Grand Island Biological Company or Microbiological Associates.

Experiments designed to look for growth-promoting conditions and factors were set up on at least two different occasions with three to four foreskins each time and a minimum of eight explants for each condition. Specific agents [i.e. insulin, epidermal growth factor (EGF), hydrocortisone, fibroblast growth factor (FGF), biotin, and 12-*O*-tetradecanoylphorbol 13-acetate (TPA)] were added to the culture medium at 24–48 hours after explanation and replenished at each medium change. Explant growth was quantitated by measurement of the diameter of the outgrowth in 70% ethanol-fixed dishes. Giemsa stain was used for photographic purposes.

Histological sections of fresh foreskins and 1-week-old explants were prepared by glutaraldehyde fixation, Epon embedding, and staining of 1-μm-thick sections with toluidine blue.

Autoradiography was performed by the addition of either [^{3}H]thymidine (dThd) or [^{14}C]dThd to the culture medium for specified time periods. After several PBS washes and Formalin fixation, Kodak NTB2 emulsion was poured into the dish and allowed to expose for 10 days prior to development.

III.   Results

A.   Tissue

A convenient feature of foreskins is that they can be stored for up to a week and probably longer prior to explantation. No difference in lag time, growth rate, or extent of outgrowth due to storage has been noted. Whole-thickness and split-thickness pieces of adult skin, stored refrigerated in moistened gauze for up to 45 days, have produced outgrowths. In general, more extensive outgrowth has been obtained with foreskin than with adult skin, although the different procedures used to obtain the two tissues may be a contributing factor.

B. Culture Preparation

Various positions and arrangements of the explants have been tried. Positioning the explant block on the dish either dermal side down or sideways appears to make no difference in the quality or extent of epidermal outgrowth, because although firm attachment to the dish is imperative, the dermal-side-down position yields more dependable cultures. Placement of the block in the upside-down position is unsuitable because adhesion to the dish is poor. Epiboly (encasement of the block by growing epidermal cells) is more complete in the epidermal-side-up position.

The extent of drying of the explant on the dish prior to the addition of medium is critical, but is unfortunately a subjective measurement that varies from one skin to another. Sufficient drying is required for good ahesion to the dish, but further drying appears to affect the dermal part of the block and can effectively inhibit fibroblast outgrowth.

The number and arrangement of the explants in the dish are determined primarily by the type of experiment and by convenience. For short-term experiments (2-3 weeks) in which size of outgrowth is used as a measure of growth or toxicity, a 2 × 2 arrangement in a 60-mm dish with the blocks 2 cm apart has been found to provide sufficient room for outgrowth. Because the outgrowth from one explant can (but does not always) eventually fill the entire dish, one or two explants per dish are more economical when large numbers of epidermal cells are needed. For long-term culture experiments, which are often beset with contamination problems, the establishment of explant cultures in flasks is suggested.

C. Characteristics of Explant Cultures

Microscopic examination of human skin explants has shown that epithelial cells begin to grow out from the tissue block within several days after explantation. This occurs by a process known as epiboly, during which the epidermal cells grow down the side of the block and out onto the dish, effectively sealing off the block. At this time the structure of the blocks is still intact, with the surface retaining its normal morphological appearance and outer covering of keratin (Fig. 1A and B). Radial outgrowth of the epithelial cells progresses with time, with the leading edge of cells giving a slightly scalloped appearance (Fig. 1C). The outgrowth is made up of cells approximately 50 μm in diameter with distinctive oval nuclei and one to six prominent nucleoli.

After the outgrowth reaches several millimeters, stratification occurs, beginning immediately around the explant and expanding outward with the growing culture so that a monolayer exists only near the leading edge (Fig. 1C). It has

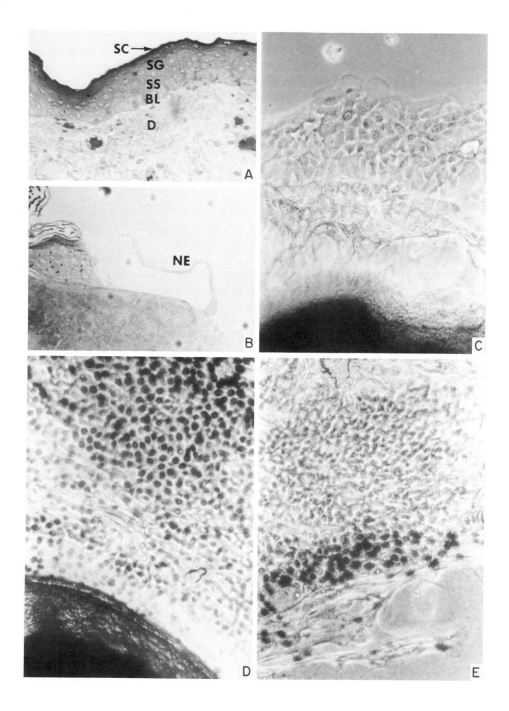

been shown (Flaxman, 1974) that cells in the lowest layer, which give the culture its cobblestone appearance, have the features of immature keratinocytes (i.e., they undergo mitosis and contain desmosomes and bundles of tonofilaments). Cells in the upper layers, which are large, flattened, and semitransparent, have the characteristics of fully differentiated cells (i.e., they lack nuclei and organelles such as ribosomes and mitochondria and contain fibrous keratin protein).

By about 10 days the original block of skin begins to degenerate, with the upper layers sloughing off into the medium. The outgrowth maintains a cobblestone appearance and usually continues to grow outward until the dish is confluent. Unlike many other cell types, the sheet of epidermal cells will continue to grow up the sides of the culture dish to slightly above the level of the medium. An interesting feature of large outgrowths that occur after several weeks is the development of ridges which appear as swirling patterns. This ridge formation was observed by Green and Thomas (1978) in nearly confluent dispersed cell cultures. They noted further that with time these ridges developed the distinctive dermatoglyph patterns of arches, whorls of either handedness, and triradials.

Once an outgrowth is well established (i.e., 2 mm or more), it is possible to remove the explant with forceps and scapel and replant it in a new dish. This process can be repeated several times with the establishment of healthy outgrowths each time. The original outgrowth will continue to grow and stratify in a radial direction.

The outgrowth of fibroblasts is a serious disadvantage in using the explant technique for long-term experiments. The incidence of fibroblast outgrowth can be decreased by several techniques, however. The use of a sharp blade and a slicing rather than a mashing motion when explants are cut from the skin will inhibit fibroblasts. This is probably related to the wounding effects seen when tissues are mashed. As mentioned above, allowing the block to dry sufficiently on the dish permits the epidermal cells to grow down the block, sealing it off to fibroblasts. If fibroblast outgrowth is troublesome, it can be controlled by careful trypsinization, since the epidermal cells are fairly resistant to trypsin, and/or by gentle scraping of the affected area with an appropriately trimmed rubber policeman.

FIG. 1.   (A) Light micrograph of a cross section of human foreskin showing normal epithelial stratification. The basal layer (BL) is over the dermis (D) and gives rise to the multilayered stratum spinosum (SS), stratum granulosome (SG), and stratum corneum (SC). ×220. (B) Cross section of a growing explant culture showing the cut edge of the epidermis. New epidermal (NE) growth extends over the block and out onto the dish. ×220. (C) Explant culture of epidermal cells. The cell sheet forms a monolayer at the leading edge but is multilayered close to the explant, which appears black at the bottom of the picture. ×220. (D and E) Autoradiographs of an explant culture showing label distribution after a 12-hour labeling with [³H]dThd. ×140.

D. Autoradiography

Mitoses can be observed microscopically throughout the radial width of the outgrowth, as has been previously noted by others (Flaxman *et al.*, 1967). This was confirmed by autoradiography of cultures labeled for 12–18 hours (Fig. 1D). [^{14}C]dThd is probably the more suitable isotope for explant cultures, since there is some indication that the lower beta energy of [^{3}H]dThd may not penetrate the multiple layers of cells sufficiently to expose the emulsion. As shown in Fig. 1D and E, autoradiography has limitations as a quantitative measure of growth because division does not always occur evenly across the outgrowth.

E. Subcultures

The use of 0.25% trypsin in 0.04% EDTA for 5 minutes at 37°C has been found to dissociate the epidermal cell sheet into a monocellular suspension. With slow centrifugation, much of the horny material remains suspended; plating of the pellet results in better than 90% cell attachment. These cultures do not regain their cobblestone appearance; rather, senescence becomes pronounced, and after several days the cells commence sloughing. If a lethally irradiated 3T3 feeder layer is used for replating the cells, a small percentage will establish growing colonies of epidermal appearance, as described by Rheinwald and Green (1975).

F. Culture Conditions

The two media used, SM and BME, represented highly enriched medium and minimal medium, respectively. The enriched medium was developed for mouse epidermal cell culture, where it was superior to the minimal medium (unpublished results). As shown in Table I, however, SM does not enhance explant outgrowth and, although variation exists from individual to individual, tends to produce slightly less outgrowth than that seen with BME. No morphological differences were apparent between cultures grown in the different media.

As indicated in Fig. 2, FCS is required for growth. On rare occasions some explants will initially show outgrowths of 1–2 mm. Maximum outgrowth was obtained with 10% or more FCS; however, the improvement seen with 15% over 10% was too small to warrant routine use of FCS concentrations greater than 10%. Although the data are not presented here, it has been found that the particular lot of FCS can be an important factor in determining the extent of outgrowth. Some lots result in 3-week outgrowths five times larger than those seen with other lots.

Incubation temperature was determined to be one of the most critical factors in the optimization of explant outgrowth. Whereas mouse epidermal cells preferred 31°C over 37°C (unpublished data), human epidermal cells showed a much reduced growth at the lower temperature, as indicated in Fig. 3.

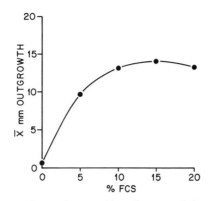

FIG. 2. FCS requirement for maximum epidermal outgrowth from explants. Three-week-old outgrowths of foreskin explant cultures grown in SM at 37°C and 5% carbon dioxide. Each point represents the mean diameter of outgrowths from explants of eight different foreskins.

G. Hormones and Growth Factors

Insulin and hydrocortisone are the two most prevalently used and efficacious hormones for many cell types. Insulin-supplemented medium, however, has an inhibitory effect on explant outgrowth, as shown in Fig. 4A. This effect is noticeable at a concentration of 1 μg/ml and is more pronounced at 10 μg/ml, the concentration used in many media preparations including SM. This may explain in part the decreased growth observed in SM as compared to BME.

As indicated in Fig. 4B, 10 ng/ml hydrocortisone increased the size of explant outgrowth by 50%. Higher concentrations were not tried, and they may prove to be even more stimulatory.

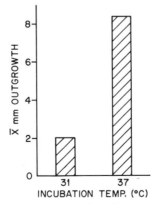

FIG. 3. Effect of incubation temperature on epidermal outgrowth from explants. Three-week-old outgrowths of foreskin explants cultured at either 31° or 37°C. Six foreskins were used in parallel cultures grown at the two temperatures.

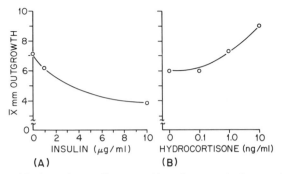

FIG. 4.   Insulin and hydrocortisone effects on epidermal outgrowths from explants. Outgrowth size from foreskin explants grown in BME plus 10% FCS. (A) Medium supplemented with 0, 1, or 10 μl/ml bovine insulin. Values are means from nine foreskins grown for 2½ weeks. (B) Medium supplemented with 0, 0.1, 1.0, or 10 ng hydrocortisone/ml. Values are means from five foreskins grown for 2 weeks.

Several specific growth factors were tested for their ability to encourage large explant outgrowths. EGF had a definite dose–response stimulatory effect in most of the foreskins studied. Figure 5 illustrates a typical example of this; the outgrowth at 20 ng/ml is twice the diameter seen in the untreated control.

FGF, as indicated in Fig. 6, had no effect on outgrowth size at doses of 5, 10, and 20 ng/ml. No increase in fibroblast contamination was observed.

The use of a lethally irradiated 3T3 feeder layer may be considered under the category of growth factors, since the improved growth seen in many systems appears to be due to a labile factor given off by the 3T3 cells. It has been reported by Rheinwald and Green (1975) that a 3T3 feeder layer is required for the growth of dispersed human epidermal cells. For this reason it was surprising to find that the use of 3T3 cells depressed the outgrowth of epidermal cells in explant culture, as shown in Fig. 7. Some preliminary studies (data not shown) have suggested that this effect is independent of medium type, and less outgrowth was observed with 1×10^6 irradiated 3T3 cells per dish than with 2.5×10^5 cells.

H.   Growth Promoters

A particular use of explant cultures is illustrated in Table II, i.e., determination of the toxicity and/or mitogenic characteristics of particular pharmacological or xenobiotic agents. In this example the potent tumor promoter TPA was used at five log dose concentrations. Unlike the response seen with mouse epidermal cells, for which a fivefold increase in DNA synthesis occurred with 0.1 μg/ml, the human explant cultures treated continuously for 3 weeks showed a decrease in growth in a dose–response manner.

I.   Quantitation of Growth and Sample—Sample Variation

Direct measurement of the size of the outgrowth is probably the easiest and best method for quantitating growth. Other methods are less suitable for the following reasons: (1) Determination of cell number following trypsinization may or may not include those cells constituting the sloughing upper layers. (2) Protein or DNA content may be inaccurate for the same reason. (3) Specific activity measurements, that is, uptake of radiolabeled precursors, are based on the above-mentioned parameters and problems. (4) Autoradiographs of explant cultures show uneven distribution of label, requiring scoring of the entire outgrowth for accurate analysis.

The extent of outgrowth under any one set of conditions has been observed to

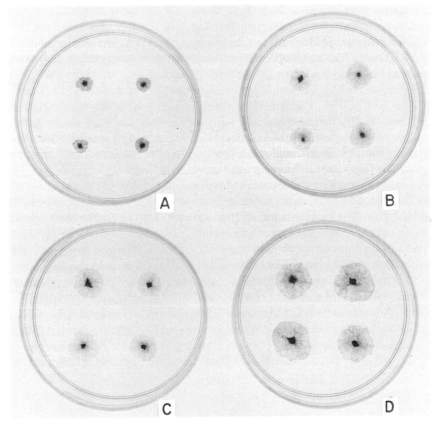

FIG. 5.   Effect of EGF. Explant cultures (2½ weeks old) grown in the presence of EGF. (A) Control; (B) 5 ng/ml; (C) 10 ng/ml; (D) 20 ng/ml.

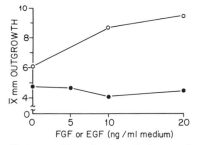

Fig. 6.   Growth factor effects on epidermal outgrowth from explants. Outgrowths (2½ weeks old) from foreskin explants grown in BME plus 10% FCS and supplemented with either 0, 5, 10, or 20 ng FGF/ml (solid circles) or 0, 5, or 10 ng EGF/ml (open circles). Values represent means from six and eight foreskins, respectively.

vary considerably from one foreskin to another, and the response to a particular agent may result in even greater differences (e.g., see Tables I and II). This provides a situation in which the data are better analyzed in terms of trends or generalizations. The large differences between individuals preclude use of the commonly employed parametric statistics.

IV.   Discussion

The usefulness of *in vitro* systems as whole-animal models depends to a large extent on the ability to retain, in culture, as many *in vivo* characteristics as possible. In the case of skin an unusual circumstance makes this more difficult than for most other tissues. A special relationship exists between the epidermis

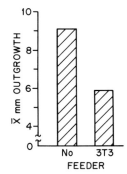

Fig. 7.   Effect of a 3T3 feeder layer on epidermal outgrowth from explants. Outgrowths (2–3 weeks old) from foreskin explants grown in the absence or presence of 1×10^6 or 2.5×10^5 lethally irradiated 3T3 cells. Values are the means of cultures from 11 different foreskins grown in either BME or SM.

TABLE I

EFFECT OF MEDIUM ON EPIDERMAL OUTGROWTH
FROM EXPLANTS[a]

Foreskin number	Outgrowth diameter (mm)	
	SM	BME
111	3.5	6.4
112	8.3	8.1
106	7.1	7.5
104	5.0	7.2
105	5.5	5.0
107	6.3	6.0
102	3.7	6.0
103	6.8	10.0
115	3.7	4.0
122	2.4	3.7

[a] Twenty-day outgrowths of foreskin explant cultures grown in either SM or BME with 10% FCS at 37°C and 5% CO_2. The values are the means of three cultures of four explants each.

TABLE II

THE EFFECT OF TPA ON EXPLANT OUTGROWTH SIZE[a]

Foreskin number	Outgrowth diameter (% of control)				
	0.001 μg	0.01 μg	0.1 μg	1.0 μg	10 μg
211		67	67	0	0
212		83	63	25	16
213		47	49	65	14
214		39	44	33	15
215		104	139	41	22
216		97		90	4
217			49	42	
218	62	87	75	63	
219	83	73	71	65	
220	49	59	49	57	
221	91	89	84	63	
$\bar{X}$	71	75	69	50	12

[a] Three-week outgrowths of foreskin explant cultures grown in either SM or BME supplemented with 0.001, 0.01, 0.1, 1.0, or 10 μg TPA/ml medium. Values represent mean outgrowth diameters from at least two dishes.

and dermis; since circulation is confined to the dermis, all nutrients, gases, and other factors must pass through the dermis and basement membrane before reaching the epidermis. The dermis, therefore, acts as a screen, as well as possibly contributing materials of its own that may regulate the growth and differentiation of the epidermis. There are indications that this phenomenon is particularly important in such conditions as wound healing (Parshley and Simons, 1950) and psoriasis (Flaxman, 1972), in which it has been suggested that the fibroblasts exert a stimulatory effect on epidermal growth. The influence of the mesenchyme on epidermal proliferation and differentiation has been explored by Wessells (1964) in a thought-provoking report on experiments with chick epidermis. He has demonstrated that the extent and nature of the differentiative pattern are dependent on mesenchymal type, age, and the vitamin or hormone environment.

The explant method of culture provides a situation that more closely approximates the whole skin than do cultures of dispersed cells. This method precludes enzymatic treatment such as trypsinization, which may alter the cell surface sufficiently and thereby be responsible for the poor plating efficiency seen in dispersed cell cultures (Flaxman, 1972). The use of explants allows the cells to retain their cell–cell contact and communication. The importance of these contacts is suggested by the observation that in explant outgrowths the epidermal cells have never been observed to emigrate independently from the cell sheet (Karasek, 1966), while fibroblasts are often seen as isolated single cells. It has been proposed that the key to the problem lies in the necessity for maintaining cell attachment. In explant cultures, growing epidermal cells are always attached to something besides each other, be it the side of the tissue block or the culture dish (Flaxman, 1972).

The spatial relationship of the epidermal cells to the other skin components, which is maintained at the time of explantation, may be an important factor in initiating outgrowth. It has been reported (Flaxman, 1974; Freeman *et al.,* 1976; Karasek, 1966; Prose *et al.,* 1967) that migration of cells from the tissue block accounts for little if any of the explant outgrowth; rather, most of the cell sheet represents new growth. As shown in Fig. 1D and as reported previously (Flaxman *et al.,* 1967; Karasek, 1966; Parshley and Simmons, 1950), mitosis continues to occur throughout the radial width of the outgrowth, although it is confined to the lower or basal layer of cells. This is analogous to the skin *in vivo* where the basal cell layer constitutes the dividing cell population of the epidermis. The upper cells in both cases represent the more mature differentiated forms. Thorough ultrastructural studies comparing maturation *in vitro* with that *in vivo* have been reported (Flaxman *et al.,* 1967; Friedman-Kein *et al.,* 1966; Liu and Karasek, 1978; Prose *et al.,* 1967). In brief, it has been shown that the nature of the cytoplasmic membrane and cytoplasmic contents of the basal layer of the explant outgrowth are similar to those of immature basal cells of the skin.

Cells of the upper layers in culture have the same features as cells in the upper layers of the skin, i.e., loss of organelles, membrane thickening, granulation, and eventual keratinization.

Attempts have been made to quantitate the number of cell generations in an explant culture (Freeman *et al.*, 1976). A major difficulty exists in determining the number of cell layers formed and the number of keratin layers that have been sloughed. Since mitosis occurs only in the basal cells, cell multiplication does not occur in a geometric progression. The flattened keratinized cells are considerably larger than the underlying basal cells. It has been estimated that only 1 in 10 basal cells needs to divide to form a layer of keratinized cells. On this basis it has been calculated that a 300-fold increase in outgrowth areas represents 13 generations (Freeman *et al.*, 1976).

Flaxman *et al.* (1967) has noted that, whereas it was previously thought that differentiation in the skin occurred as a result of increased distance from the supply of nutrients, the situation in culture is reversed; i.e., the most mature nonproliferating cells have the most contact with the culture medium. Therefore, the determinant of the proliferation–maturation gradient is attachment of the basal cells to a suitable surface. This view is further supported by the observation (Flaxman *et al.*, 1967; Flaxman, 1972) that the basal cells of detached sheets of the outgrowth all mature, with cessation of proliferation. It has been suggested that attachment itself maintains the cells in an undifferentiated state and that the basement membrane serves this function *in vivo*. The importance of the nature of the substrate to the structural characteristics of various epithelial cells has been investigated in human cells by Reinertson (1961) and in chick skin by Wessells (1964).

Many of the distinctive properties of the skin are a direct consequence of the differentiation of epidermal cells into keratinized cells. The keratinization process, however, is not uniform (Prose *et al.*, 1967). There are many similarities between the keratin layers in explant cultures and those in skin, including keratinization of the upper but not lower layers, abruptness of the transition from nonkeratinized to keratinized cells, thickened outer cell membranes in keratinized cells, and lack of desmosomes between keratinized cells (Friedman-Kein *et al.*, 1966; Prose *et al.*, 1967). However, differences between keratinized cells of outgrowths and of skin also exist. *In vivo* maturation of the epidermis is marked by the appearance of keratohyalin granules and the concurrent formation of stratum corneum. Some controversy exists as to whether true stratum corneum and keratohyalin granules are present (Flaxman, 1972) or not (Flaxman *et al.*, 1967; Prose *et al.*, 1967) in explant cultures, although distinctive granules nearly always appear in the upper layers (Flaxman, 1974; Prose *et al.*, 1967). In skin during keratinization, cytoplasmic elements and nuclei are quickly decomposed, but in the explant outgrowth the process is often less efficient and incomplete, resulting in occasional retention of the nucleus (Prose *et al.*, 1967). This process of proliferation–maturation does not depend on the presence of dermal elements in that, once the epidermal cell sheet has grown out onto the culture dish, the explant can be removed without continued outward growth and differentiation being altered.

The "purity" of epidermal outgrowths has been well described (Flaxman *et al.*, 1967; Karasek, 1966; Prose *et al.*, 1967), and it is commonly accepted that, while fibroblast contamination occurs, it is as distinctive areas of growth outside the epidermal cell sheets. Neither fibroblasts nor collagen has been detected in an epidermal cell sheet. Fibroblasts display unique distinguishing characteristics: They grow only as monolayers when confluent; the cells are often separated by large distances and show few cell contacts, even when crowded; and they synthesize collagen but not keratin (Flaxman, 1974). The involvement of cells other than fibroblasts in the epidermal outgrowth has been investigated. Melanocytes do not appear to be included (Prose *et al.*, 1967), nor do Langerhan's cells (Prunieras, 1969), although a mobile but nonreplicating small cell resembling a lymphocyte has been reported (Flaxman, 1974). The explant method, therefore, allows a high degree of confidence that relatively large populations of epidermal cells can be obtained in which fibroblasts are absent.

Although it is an easy task to maintain skin fragments for relatively long periods of time (up to 4 weeks or more), and only minimal medium is required to initiate an outgrowth of epidermal cells, continued proliferation and organized maturation are affected by various environmental factors. These include specific nutritional and hormonal requirements, growth factors, incubation temperature, pH, gas tensions, and type of substrate. A detailed review of the methodology employed in the numerous past attempts to culture skin will be avoided. Rather, more recent and definitive work, especially the inclusive study by Karasek (1966), will be discussed.

As with many other cell types, the choice of medium is an important factor in optimization of culture conditions. This problem has been studied in some detail (Karasek, 1966). Growth rates in media of varying degrees of complexity were observed, and none was found to be superior to Eagle's minimal essential medium (MEM); the slighly inhibitory effect of highly enriched media was also observed in this study. A new medium, NCTC 168, was recently developed specifically for human skin explants (Price *et al.*, 1978). Supplementation of the medium with additional L-glutamine has been found to be beneficial (Karasek, 1966; Reinertson, 1961). An absolute requirement for the sulfur amino acid cysteine and a growth-promoting effect for methionine were noted by Karasek (1966). In general, most studies have been done with Eagle's MEM (Flaxman and Harper, 1975a,b; Freeman *et al.*, 1976; Friedman-Kein *et al.*, 1966; Pinkus, 1959; Reinertson, 1961). It is possible that the better growth observed in less complex media is the result of an evolutionary adaptation by the epidermis to the dermal filtering system. It was also noted in this study that the more enriched medium, SM, supported earlier and more extensive fibroblast contamination.

The use of medium supplemented with 10% FCS has been standard practice for tissue culture work in general, and explant cultures are no exception. The effect of serum on the growth of explants has been the subject of several studies. Karasek (1966) found that concentrations of calf serum higher than 10% did not appreciably increase growth rates. Furthermore, at equivalent concentrations, it was found that pooled adult human serum decreased the outgrowth by half over that seen with calf serum. Reinertson (1961) observed very poor growth in calf serum alone (no medium) when compared with Eagle's MEM with 10% serum. The data reported here suggest that, in general, 10% FCS is sufficient, but for some individuals 15% results in a significantly greater growth rate. An absolute requirement for serum was also noted, although some foreskins gave small (1-mm) early outgrowths in the absence of serum.

Other culture conditions shown to affect outgrowth are pH, oxygen tension, and incubation temperature. The pH of the medium needed for maximal epithelial outgrowth was reported by Karasek (1966) to be 7.2–7.4, with more alkaline pH (>7.8) or more acid pH (<7.0) decreasing growth markedly. Others (Price et al., 1978) have found, however, that a medium near neutral pH or slightly acidic promotes better growth.

It has been shown by several investigators that epidermal outgrowth is best at oxygen tensions of 20–40%, while it is markedly inhibited by high tension (95%) and does not occur at tensions under 5% (Karasek, 1966). It was also observed by Taylor et al. (1978) that the use of 10% oxygen retarded epithelial growth but enhanced fibroblast growth rate and colony development.

Very little work has been reported on the effect of different incubation temperatures. As shown in this report, the use of a 31°C incubation markedly inhibited growth. This is in contrast to work with other species, especially murine epidermal cells, in which 31°C was found to promote culture growth and longevity.

In recent years it has become common practice to supplement standard medium formulations with various hormones, vitamins, and growth factors specifically beneficial to the tissue of interest. While human epidermal cells do not require these additions for moderate growth, stimulatory effects have been observed with specific factors.

EGF is a polypeptide that has been found to enhance growth and maturation of epidermis in newborn mice, organ cultures of epithelial tissues (Rheinwald and Green, 1977), and cultures of dispersed human keratinocytes (Peehl and Ham, 1978; Rheinwald and Green, 1975 and 1979). In a short report on growth requirements of dispersed keratinocytes, Peehl and Ham (1978) noted that EGF stimulated growth and increased the longevity of the cells only when added in the presence of fibroblasts. In the explant method of growing epidermal cells, EGF was found to be very effective in stimulating growth. This was evident in cultures several weeks old in which the original tissue blocks, although still present, had almost entirely degenerated. FGF has been found to be ineffectual in dispersed cell cultures (Peehl and Ham, 1978), as well as in explant cultures (Section III,F).

The use of a lethally irradiated feeder layer has been shown to be required for the initiation of colony formation and for continued growth (Sun and Green, 1976) in dispersed epidermal cell cultures. Flaxman (1972) tried air-dried monolayers of fibroblasts, since he noted that explants would not stick to a wet substrate. This approach did not produce any significant difference in the rate of growth. The method tried here, plating 3T3 cells around explants that had already adhered to the dish, significantly decreased the outgrowth. There are no obvious explanations for this; it is unlikely that the 3T3 cells exert their negative effect directly. While it is possible that the 3T3 cells are depleting the medium of a particular nutrient or otherwise altering it, this effect was seen with as few as 2.5×10^5 3T3 cells per dish, and this does not seem to be the case with dispersed cells.

The hormone requirements of cultured skin have not been well defined. Insulin, although used in many media formulations at concentrations of $1-50 \mu g/ml$ for growth enhancement, appeared to inhibit explant growth, even at $1 \mu g/ml$. Hydrocortisone, on the other hand, has been shown to be a beneficial addition to media ($0.4 \mu g/ml$) for dispersed human epidermal cell cultures grown on a 3T3 feeder layer (Sun and Green, 1978). Peehl and Ham (1978) found that high levels of hydrocortisone ($10 \mu g/ml$) or dexamethasone were needed for the growth of dispersed cell cultures in unconditioned medium. The study reported here suggests that, while hydrocortisone is not required for the growth of explant cultures, it has a growth-promoting effect in foreskin cultures.

Of the vitamins the most widely studied with regard to skin is vitamin A. In a study on growth and differentiation of epidermal cells in explant cultures, Chopra and Flaxman (1975) showed that daily 30-minute exposure to vitamin A alcohol markedly decreased the number of keratohyalin granules without evidence of mucous metaplasia. In addition, the treated outgrowths were larger and had a higher mitotic index.

Of great interest was the response of foreskin outgrowths to the tumor promoter TPA. One of the characteristics of tumor promoters was stimulation of epidermal proliferation in the whole animal and increased DNA synthesis in epidermal cells in culture (Slaga *et al.*, 1978a). These studies were carried out primarily in the mouse, and it is of obvious interest to determine if the mouse is a suitable model for human skin. The results of this experiment suggested that the majority of the individuals tested showed a toxic response, with depressed DNA synthesis at doses that normally stimulate in the mouse. It is possible that the use of foreskins is inappropriate. Experiments are in progress to determine if adult skin responds similarly.

V.   Perspectives

Cultures of human epidermal cells have many far-reaching uses. The continuous proliferation–maturation phenomenon of epidermis makes it an ideal material for inquiries into regulatory mechanisms in differentiation processes. The ease of explant methodology offers the possibility of the development of a human test system for studying skin responses to various pharmacological or xenobiotic agents. Individual characteristics of growth and response to such agents present the opportunity for investigations into the range of human variability and human genetics.

The most useful feature of human skin culture, however, lies in carcinogenesis studies. It has been established that the majority of human cancers are epithelial in origin. Skin is, therefore, not only particularly relevant, but also the most readily accessible epithelial tissue. Explant cultures may be the best model for carcinogenesis studies, because the normal architecture of the skin is maintained. There are several lines of evidence that suggest the importance of retaining normal skin relationships, especially at the time of carcinogen exposure.

In a review on the dermal response to carcinogens, Pinto et al. (1968) point out through examples from various species that cutaneous carcinogenesis is nearly always characterized by alterations in the dermis. Since it has been demonstrated that dermal changes may exert a considerable, if not critical, influence on epidermal carcinogenesis (Pinto et al., 1968), maintenance of dermoepidermal interaction through the use of an explant rather than a dispersed cell culture may be required for the in vitro transformation of human epidermis. It has been suggested that one explanation for the difficulty found in transforming human diploid cells is the lack of suitable culture conditions.

Maintenance of cells retaining their differentiated characteristics, as seen in the explant outgrowth, may be important in determining the tumor type and differentiation. In studies on the transformation of cultures of dispersed cells from newborn mice, Slaga et al. (1978b) noted that such cultures lost their epithelial characteristics shortly after plating and that the tumors resulting from carcinogen treatment of such cultures were of a highly undifferentiated nature. In contrast, transformation of trachea epithelium by exposure to carcinogens in organ culture, followed by explantation to allow the epithelial cells to grow out, resulted both in outgrowths of well-differentiated cells and in the formation of well-differentiated squamous cell carcinomas (Steel et al., 1978).

Explant cultures of genetically cancer-prone and -resistant individuals may be a fruitful approach to elucidation of the mechanisms of carcinogenesis. It is well known that there are several genetically determined diseases such as xeroderma pigmentosum, Fanconi's anemia, and Bloom's syndrome in which individuals have an increased if not invariable incidence of cancer. These diseases, which are associated with DNA repair defects, may offer a steppingstone, particularly with regard to the tumor promotion stage of carcinogenesis. Promotion is probably the rate-limiting factor in cancer, since the population at large is continually barraged with environmental carcinogens.

Another area needing further exploration is the difference between newborn and adult tissues. Studies on neonatal and adult mice have shown that they respond in a nearly inverse manner to the tumor promoter TPA. Developmental differences in receptors and metabolic capabilities most likely exist.

VI. Summary

Human skin epithelial cells grown *in vitro* from explants display the same organizational behavior found *in vivo*. Culture conditions found to encourage epidermal outgrowth from explants include minimal rather than complex media and supplementation of such media with 10% FCS, EGF, and hydrocortisone. The explant method provides a means of obtaining a large number of keratinocytes from a small piece of tissue. These primary cultures are useful in studies on the proliferation–differentiation process in skin and on the response of skin to various pharmacological agents, and *in vitro* transformation studies in which explants are potentially the best model for human carcinogenesis.

References

Chopra, D. P., and Flaxman, B. A. (1975). *J. Invest. Dermatol.* **64,** 19–22.

Flaxman, B. A. (1972). *In Vitro* **8,** 237–256.

Flaxman, B. A. (1974). *In Vitro* **10,** 112–118.

Flaxman, B. A., and Harper, R. A. (1975a). *J. Invest. Dermatol.* **64,** 96–99.

Flaxman, B. A., and Harper, R. A. (1975b). *Br. J. Dermatol.* **92,** 305–309.

Flaxman, B. A., Lutzner, M. A., and VanScott, E. J. (1967). *J. Invest. Dermatol.* **49,** 322–332.

Freeman, A. E., Igel, H. J., Herrman, B. J., and Kleinfield, K. L. (1976). *In Vitro* **12,** 352–362.

Freidman-Kein, A. E., Prose, P. H., Liebhaber, H., and Morill, S. (1966). *Nature (London)* **212,** 1583–1584.

Green, H., and Thomas, J. (1978). *Science* **200,** 1385–1388.

Karasek, M. (1966). *J. Invest. Dermatol.* **47,** 533–540.

Liu, S.-C., and Karasek, M. (1978). *J. Invest. Dermatol.* **71,** 157–162.

Parshley, M. S., and Simms, H. S. (1950). *Am. J. Anat.* **86,** 163–189.

Paul, J. (1961). *In* "Cell and Tissue Culture" (O. Ljunggren, ed.), p. 1. Williams & Wilkins, Baltimore, Maryland.

Peehl, D. M., and Ham, R. G. (1978). *J. Cell Biol.* **79,** 78.

Pinkus, H. (1959). *J. Invest. Dermatol.* **33,** 171–175.

Pinto, J. S., Dobson, M. C., and Dobson, R. L. (1968). *Adv. Biol. Skin* **10,** 195–220.

Price, F. J., Camalier, R. F., and Gantt, R. R. (1978). *Tissue Cult. Assoc. Meet.* Abstr. 76, p. 251.

Prose, P. H., Friedman-Kein, A. E., and Neistein, S. (1967). *Lab. Invest.* **17,** 693–716.

Prunieras, M. (1969). *J. Invest. Dermatol.* **52,** 1–17.

Pullman, H., Lennartz, K. J., and Steigleder, G. K. (1974). *Arch. Dermatol. Forsch.* **250,** 177–184.

Reinertson, R. P. (1961). *J. Invest. Dermatol.* **36,** 345–352.

Rheinwald, J. G., and Green, H. (1975). *Cell* **6,** 331–344.

Rheinwald, J. G., and Green, H. (1977). *Nature (London)* **265,** 421–424.

Slaga, T. J., Fischer, S. M., Viaje, A., Berry, D. L., Bracken, W. M., LeClerc, S., and Miller, D. R. (1978a). *In* "Mechanisms of Tumor Promotion and Carcinogenesis" (T. J. Slaga, A. Sivak, and R. K. Boutwell, eds.), pp. 173–202. Raven Press, New York.

Slaga, T. J., Viaje, A., Bracken, W. M., Buty, S. G., Miller, D. R., Fischer, S. M., Richter, C. K., and Dumont, J. N. (1978b). *Cancer Res.* **38,** 2246–2252.

Steele, V. E., Marchok, A. C., and Nettesheim, P. (1978). *In* "Mechanisms of Tumor Promotion and Carcinogenesis" (T. J. Slaga, A. Sivak, and R. K. Boutwell, eds.), pp. 289–300. Raven Press, New York.

Sun, T.-T., and Green, H. (1976). *Cell* **9,** 511–521.

Sun, T.-T., and Green, H. (1978). *J. Biol. Chem.* **253,** 2053–2060.

Taylor, W. G., Camalier, R. F., and Sanford, K. K. (1978). *Tissue Cult. Assoc. Mett.* Abstr. 74, p. 352.

Wessells, N. K. (1964). *J. Exp. Zool.* **157,** 139–152.

Chapter 15

Serial Cultivation of Normal Human Epidermal Keratinocytes

JAMES G. RHEINWALD[1]

Laboratory of Tumor Biology,
Sidney Farber Cancer Institute, and
Department of Physiology,
Harvard Medical School,
Boston, Massachusetts

[1]The author's investigations have been supported by grants from the National Cancer Institute and the Massachusetts Division of the American Cancer Society.

I.   Introduction

Until rather recently, the connective tissue fibroblast was the only normal diploid human cell type that could be serially cultivated for many cell generations. Conclusions drawn from *in vitro* studies of growth regulation, mutagenesis, carcinogenesis, replicative senescence, and the expression of differentiated function in the fibroblast cannot necessarily be generalized to all vertebrate cell types. The need for comparative studies, particularly with cells of epithelial tissues, has long been recognized.

Fibroblasts can be grown from almost any explanted human tissue, and can be serially passaged and even cloned from low-density platings because they grow well, quite gratuitously, in several media originally developed for the culture of established cell lines of other species. Epithelial cells from most tissues do not fare nearly as well in monolayer culture in these media, and most attempts to produce even primary epithelial cell cultures of any tissue have resulted in overgrowth by stromal fibroblasts. Primary cultures of epidermal keratinocytes had been successful using standard culture media and methods, but subcultivation and clonal growth were not achieved until recently.

In 1974, while isolating epithelial cell lines from the differentiated portion of a transplantable mouse teratoma, it was found that one epithelial cell type would grow only when cocultivated with fibroblasts. γ-Irradiated 3T3 cells sufficed as the fibroblast "feeder" population for this mouse epithelial cell line, named XB, which proved to be a keratinocyte (Rheinwald and Green, 1975a). Application of the 3T3 feeder layer to the cultivation of epidermal keratinocytes from human foreskin was successful (Rheinwald and Green, 1975b) and has permitted investigation of cytoskeletal structure and the mechanisms of growth control, terminal differentiation, and replicative senescence in this cell type. The potential usefulness of cultured human keratinocytes in the biochemical and genetic analysis of functions not expressed by the more easily cultivated fibroblast is only beginning to be appreciated. These methods and perspectives are presented to encourage use of the keratinocyte culture system as a routine tool in studying the biology of human cells.

II. Materials and Methods

A. Culture Media and Solutions

Dulbecco's modification of Eagle's medium (DME) is obtained in powder form from a commercial supplier and made up in 10-liter batches in the laboratory. Then 0.1 mg/ml penicillin and streptomycin are added. The medium is saturated with an atmosphere of 100% CO_2, filter-sterilized into 100-ml bottles, and stored at 4°C for up to 1 month or at -20°C for up to 6 months.

DME is supplemented with 10% calf serum (selected lots) for the growth of 3T3 cells, or with 20% fetal calf serum (FCS) (selected lots) and 0.4 μg/ml hydrocortisone for keratinocyte cultures growing with 3T3 feeder cells. Mycostatin (Grand Island Biological Company) is added at 20 μg/ml as a fungicide to primary cultures only. Cultures are grown at 37°C in an atmosphere of 8% CO_2 and are refed twice weekly.

Mouse epidermal growth factor (EGF) is obtained from Collaborative Research, Inc., or is prepared in the laboratory by the method of Savage and Cohen (1972).

Trypsin, collagenase, and EDTA solutions are made in neutral phosphate-buffered saline (containing 8.0 gm NaCl, 0.2 gm KCl, 1.15 gm Na_2HPO_4, and 0.2 gm KH_2PO_4 per liter of water). The tetrasodium salt of EDTA (sodium versenate), crystalline hog pancreatic trypsin, and clostridial collagenase (CLS type II, Worthington Biochemical) are used. The pH of the trypsin solutions is adjusted to 7.3 with NaOH to increase activity.

3T3 conditioned medium is prepared by feeding confluent 3T3 cultures with medium plus 20% fetal calf serum, harvesting 24 hours later, and filter-sterilizing before use. 3T3 "conditioned dishes" are prepared by treating confluent 3T3 cultures with mitomycin C (as described below) and removing the cells with a 0.02% EDTA solution.

B. Preparation of 3T3 Feeder Layers

The established fibroblast line 3T3, developed from random-bred Swiss mouse embryo cells (Todaro and Green, 1963; Goldberg, 1977), is used as a feeder layer for clonal growth of human keratinocytes (Rheinwald and Green, 1975b). (Feeder layers of diploid mouse and human fibroblasts also promote human keratinocyte growth, but usually do not result in such a high colony-forming efficiency as 3T3.) Confluent 3T3 cultures are rendered reproductively dead by irradiating with 6000 rads of γ rays in a cobalt source or by incubating with 4 μg/ml mitomycin C for 2 hours. Treated 3T3 cultures are trypsinized and re-

F.   Determination of Replicative Potential (Culture Lifespan)

Primary cultures containing discrete keratinocyte colonies after 8–10 days of growth (generally dishes that had been inoculated with sufficient cells to give rise to ≤ 25 colonies/cm^2 dish surface area) are chosen for subculture. Secondary and subsequent cultures are inoculated at a range of densities from 3×10^3 to 3×10^5 keratinocytes per dish with the standard number of 3T3 feeder cells such that some dishes will yield about 25 colonies/cm^2.

At each passage, dishes not used for subculture are fixed and stained to determine colony-forming efficiency—defined as the percentage of inoculated cells that give rise to colonies >100 cells. The average number of cell generations undergone by a strain during each passage is then calculated as

$$\log_2 \left(\frac{\text{number of cells per dish when subcultured}}{\text{number of colony-forming cells that initiated the culture}} \right).$$

G.   Keratinocyte Growth in the Absence of a 3T3 Feeder Layer

1.   METHOD 1

Keratinocyte colonies are initiated by plating a single-cell suspension of keratinocytes with 3T3 feeder cells in the usual way such that less than 300 colonies arise. Seven to nine days later, the feeder layer is selectively detached with EDTA, the dish is rinsed repeatedly to remove all feeder cells, DME plus 20% FCS plus 0.4 μg/ml hydrocortisone is added back, and the cultures are returned to the incubator to permit further growth of the keratinocyte colonies.

2.   METHOD 2

A single-cell suspension of keratinocytes in 3T3 conditioned medium plus hydrocortisone is inoculated at a density of $\leq 1.5 \times 10^3$ cells/cm^2 into a standard tissue culture dish or into a dish previously "conditioned" by a confluent population of 3T3 cells (see above).

3.   METHOD 3

A single-cell suspension of keratinocytes in DME plus 20% FCS plus hydrocortisone is inoculated at a density $\geq 5 \times 10^4$ cells/cm^2 into a standard tissue culture dish.

H. Inducing Terminal Differentiation by Suspension in Semisolid Medium

Medium is made semisolid by the addition of methylcellulose (4000 cps viscosity rating) at a concentration of about 1.3% (w/v). Three grams of dry methylcellulose is autoclaved with a 1-in. magnetic stir bar in a 250-ml plastic centrifuge bottle with a screw cap, and 165 ml of serum-free medium preheated to 60°C is added and stirred at room temperature for 30 minutes to completely wet and disperse the powder. This suspension is then stirred at 4°C for 2 hours, during which time the methylcellulose makes a transparent, slightly viscous solution. Next, 41 ml of FCS is added and mixed by stirring. (Semisolid conditioned medium is prepared by making a 7.25% methylcellulose solution in Earle's salts and then diluting to 1.45% with conditioned medium.) Finally, the bottle and its contents are placed over a 5-ml water cushion in a Sorvall type GSA rotor and centrifuged at 4°C at 15,000 g for 30 minutes. Most undissolved fibrous material is pelleted, permitting the clear methylcellulose medium to be poured off as the supernatant into sterile bottles to be refrigerated or frozen until use.

Tenfold concentrated cell suspensions in culture medium are added to yield the desired cell density and a final methylcellulose concentration of about 1.3%. Keratinocytes are routinely suspended in methylcellulose medium at a density of $1–3 \times 10^5$ cells/ml and incubated in loosely capped plastic centrifuge tubes to reduce the rate of evaporation.

At various times after suspension, cells are recovered from the methylcellulose medium by diluting 15-fold with regular medium, pelleting by the usual low-speed centrifugation, and resuspending in regular medium. Cells can then be replated with 3T3 feeder cells to determine colony-forming ability or resuspended in phosphate-buffered saline for sequential treatment with 1% sodium dodecyl sulfate (SDS) [to determine the percentage of cells having a disulfide-bonded cytoplasmic keratin filament matrix (Sun and Green, 1978a; Green, 1977)] and with 1% SDS plus 20 mM dithiothreitol (DTT) followed by heating to 90°C for 5 minutes [to determine the percentage of cells that have formed a cornified envelope beneath the plasma membrane (Sun and Green, 1976; Green, 1977)]. The SDS-resistant structure and the SDS-plus-DTT-resistant structure are easily identified in a hemacytometer chamber by phase-contrast or Nomarski optics.

I. Metaphase Chromosome Preparations

A modification of a standard technique (Tjio and Puck, 1958; Rothfels and Siminovitch, 1958) is used. The 3T3 feeder layer is removed from a keratinocyte culture growing in the presence of EGF about 3 days before individual colonies would begin to contact each other. 3T3-conditioned medium plus hydrocortisone is added back, and the culture is returned to the incubator. Two days later, 10^{-6} M colchicine is added to the still exponentially growing culture. After 2 hours the medium is removed and saved in a centrifuge tube, and the culture is disaggregated to single cells with trypsin plus EDTA. The cells are pelleted by low-speed centrifugation and resuspended in a hypotonic phosphate-buffered saline solution (2.25 gm NaCl, 0.34 gm Na_2HPO_4, and 0.054 gm $Na_2HPO_4 \cdot H_2O$ per liter distilled water) prewarmed to 37°C. The suspension is incubated at 37°C for 10–20 minutes. The cells are fixed in suspension by the dropwise addition of a freshly prepared solution of cold methanol–glacial acetic acid (3:1 v/v) to a final concentration of about 20% of the volume of the cell suspension.

The cells are pelleted by low-speed centrifugation in a plastic centrifuge tube, resuspended in cold methanol–acetic acid, and held in an ice bath for 20 minutes. The pelleting, resuspension in cold fixative, and incubation at 0°C is repeated two more times. Finally, the cells are resuspended in cold fixative at a concentration of about 10^7/ml and applied dropwise to scrupulously clean microscope slides which have been chilled to 0°C and placed standing at a slant. The slides are air-dried and stained with Giemsa.

J. Cryopreservation

Exponentially growing keratinocyte cultures are disaggregated to single cells and resuspended at a concentration of 2×10^5 to 10^6 cells/ml in medium plus serum. Sterile glycerol is added to 10%, and the suspension is introduced into 1-ml glass or plastic ampules with a Pasteur pipet. The ampules are slowly brought to −190°C by placing them within a styrofoam container and exposing to the vapors at the top of a Linde liquid nitrogen freezer tank for at least 2 hours before storing in their permanent location in the freezer.

Ampules are thawed by placing in a 37°C water bath. The thawed suspension is transferred to a plastic centrifuge tube with a Pasteur pipet. Ten volumes of warm medium plus serum are slowly added dropwise. The suspension is then pelleted, resuspended in culture medium, and plated at the desired density with 3T3 feeder cells.

III.   Results

A. Keratinocyte Colony Formation in Primary Cultures of Disaggregated Skin

Newborn foreskin is a convenient source of human keratinocytes possessing a high proliferative capacity in culture. Trypsin alone is sufficient to completely disaggregate this tissue to single cells with a yield of about 3 million cells per square centimeter of epidermal surface (Table I). However, the collagen fiber matrix in the dermis of older individuals is quite refractory to trypsin digestion, and the resulting fibrous mass that forms in the spinner flask appears to trap many of the released cells. The addition of collagenase has markedly improved the disaggregation of adult skin samples, so trypsin plus collagenase is now routinely used for all tissue samples in our laboratory.

Most of the cells are released between 90 and 210 minutes of enzyme digestion. The majority of colony-forming keratinocytes are released later in this time period, while the majority of colony-forming fibroblasts are released earlier. Dissociated foreskin cells usually give rise to large keratinocyte colonies in the feeder layer system with a frequency of about 0.5% (Rheinwald and Green, 1975b; also Table I). A typical yield is on the order of 10^4 colonies in the primary cultures derived from one foreskin. Virtually all colonies that arise in the primary culture grow progressively with a doubling time of about 24 hours until the entire surface area of the culture vessel is filled with keratinocytes at a density approaching 10^7 cells/100-mm dish. Distributing the disaggregated foreskin cells among a large number of dishes could, therefore, give rise to primary cultures containing a total of more than 10^8 cells within 2 weeks of plating.

TABLE I

Yield of Colony-Forming Cells from Newborn Human Foreskin

Foreskin	Epidermal surface area (cm²)	Enzymes	Time interval during digestion (minutes)	Number of cells released	Percentage of cells forming colonies	
					Keratinocyte[a]	Fibroblast[b]
K	1.5	Trypsin	30–90	6×10^5	0.13	0.5
			90–150	2.1×10^6	0.65	1.1
			150–210	2.3×10^6	0.45	0.8
			210–270	3×10^5		
			Pooled, 30–120	5×10^6	0.6	0.6
E	2.0	Trypsin	Total, 60–180	4×10^6	0.15	0.2
G	2.5	Trypsin	Total, 45–195	7×10^6	0.5	2.0
H	2.5	Trypsin	Total, 60–180	9×10^6	0.3	1.7
C	3.0	Trypsin	Total, 90–210	9×10^6	0.08	0.2
A1	1.5	Trypsin plus	45–105	1.6×10^6	0.1	0.5
		collagenase	105–175	6.4×10^6	1.0	0.25

[a] Suspension plated in DME plus 20% FCS plus 0.4 μg/ml hydrocortisone with 2×10^4 3T3 feeder cells/cm².

[b] Suspension plated in DME plus 20% FCS with 2×10^3 3T3 feeder cells/cm², or in F10 or M199 plus 20% FCS without 3T3 feeder cells.

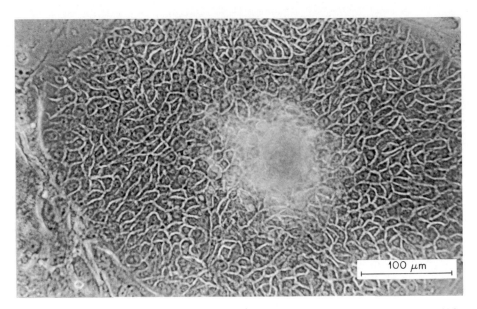

Fig. 1. A small human epidermal keratinocyte colony in primary culture, growing with 3T3 feeder cells (seen at the left at the edge of the colony). The basal layer of small, polygonal epithelioid cells is in focus, while the thick colony center and the stratified layers of very thin, flat squamous cells are not.

The keratinocytes attach directly to the culture vessel surface and push away the 3T3 feeder cells as the colonies expand. As seen in Fig. 1, the keratinocytes are closely packed in a polygonal epithelioid arrangement. The displaced feeder layer cells often appear as a rolled-up "cuff" at the colony periphery. A large number of fibroblasts with colony-forming potential are also released from the skin, but these grow extremely slowly when surrounded by a dense monolayer of 3T3 feeder cells (Rheinwald and Green, 1975b). After about a week, however, as the feeder layer density begins to decrease through cell detachment, dermal fibroblast proliferation can become a problem. The selective removal of fibroblasts with EDTA a week after plating and the addition of fresh feeder cells for the remainder of the culture period gets rid of all or nearly all dermal fibroblasts. If necessary, this operation is repeated 1 week after plating the secondary culture. Populations are easily checked for the presence of rare dermal fibroblasts by plating samples in the absence of a feeder layer with a good cloning medium for human fibroblasts such as M199 (Pious *et al.*, 1964) or F10 (Jacobs and DeMars, 1977), or with DME and a sparse 3T3 feeder layer (i.e., 2×10^3 cells/cm^2) (Rheinwald and Green, 1975b; also Table I).

B. The Behavior of Human Keratinocytes during Serial Subcultivation

Keratinocytes are optimally subcultured within 2 weeks after plating, before individual colonies have attained a size of about 10,000 cells. Small colonies are relatively easy to disaggregate to single cells with trypsin plus EDTA, while large colonies and confluent cultures are more refractory and require a much longer trypsinization time to achieve total disaggregation (Fig. 2). Secondary and subsequent cultures are initiated from low-density platings of single-cell suspensions in the same way as the primary.

A 3T3 feeder layer continues to be a requirement for the initiation of colonies from single keratinocytes at each passage. The frequency of colony formation by single keratinocytes decreases rapidly as the 3T3 feeder layer density is reduced (Fig. 3). Colony-forming frequency is increased somewhat at lower feeder layer densities if medium conditioned by confluent 3T3 cultures is used or if the feeder cells are plated 24 hours before adding the keratinocytes. At a feeder layer

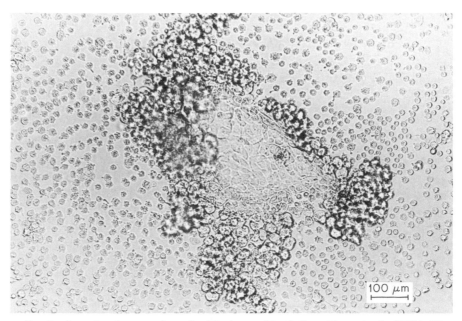

Fig. 2. Large keratinocyte colony after 15 minutes of incubation with trypsin plus EDTA. The large squamous cells of the upper layers have become detached from the basal layer cells and are retracting toward the colony center, exposing the small basal cells which are beginning to detach. Eventually, the entire colony became disaggregated into single cells.

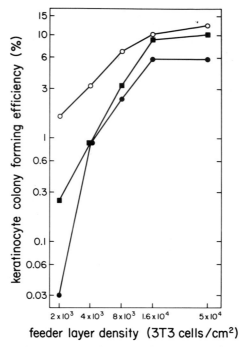

FIG. 3.   The effect of feeder layer density on keratinocyte colony-forming efficiency. Foreskin strain K cells (10^4 cells, 20 cell generations in culture at the time of plating) were plated with various densities of γ-irradiated 3T3 cells in DME plus 20% FCS plus 0.4 μg/ml hydrocortisone (solid circles), in 3T3-conditioned medium (open circles), or in fresh medium onto γ-irradiated 3T3 cells that had been plated 24 hours earlier (squares). The cultures were fixed and stained with rhodanile blue 8 days later.

density equal to that attained by 3T3 cells grown to saturation in DME plus 10% calf serum ($\cong 5 \times 10^4/cm^2$), the frequency of keratinocyte colony formation is high, but colony growth rate is slowed by the dense feeder population.

When subcultured within 10 days of plating at each passage, newborn foreskin keratinocytes can undergo a total of about 30–40 cell generations with colony-forming efficiencies in the range of 0.5–5% before they senesce, while epidermal keratinocytes from older donors generally stop growing by 25 cell generations and display a lower average colony-forming efficiency throughout their culture lifetime (Rheinwald and Green, 1975b). As described below, supplementation of the medium with EGF causes a large increase in colony-forming efficiency and replicative lifespan and is now routinely used for any culture that later will be subcultured.

Recovery of keratinocytes frozen in liquid nitrogen is very good—generally from 30 to 100%, measured as the relative colony-forming ability of suspensions before and after freezing.

C. Staining Characteristics of Keratinocyte Colonies

The traditional histological identification of keratinizing epithelia is made on hematoxylin- and eosin-stained tissue sections by the bright yellow-orange staining of the terminally differentiated squamous cells. When viewed from above, the well-stratified central portions of unsectioned keratinocyte colonies stained with hematoxylin and eosin fail to show hematoxylin staining of the underlying basal cell nuclei, perhaps because the squamous upper layers act as a barrier to the penetration of hematoxylin down through the colonies. Less stratified keratinocyte colonies, such as those growing with EGF or in the absence of a feeder layer, do not give a hematoxylin and eosin staining pattern unique to keratinocytes, however,

The rhodanile blue method of MacConaill and Gurr (1964) stains only keratinocyte colonies red and, although the color is less dense in poorly stratified colonies, the rhodamine is bound and retained by keratinocytes to a much greater extent than by fibroblasts or any other epithelial cell type (Rheinwald and Green, 1975a,b). This method is quick and easy for macroscopic or low-power microscopic examination of cultures for colony size and number, but the rhodamine does not remain tightly bound and quickly bleeds out of stained preparations during ethanol dehydration or when mounted with aqueous mounting solutions. It is therefore an unsuitable stain when examination of cell or colony structure at high magnification is desired.

The procedure of Ayoub and Shklar (1963), when applied to cultured keratinocyte colonies (Steinberg and Defendi, 1979), stains fibroblasts and unstratified regions of keratinocyte colonies blue and stratified regions yellow to red. Unlike rhodamine, these dyes remain tightly bound to the fixed material through air-drying or ethanol dehydration and mounting in xylene-based solutions. The intense orange or red staining of individual cells or small clusters of cells in a colony can be observed under the microscope. This staining system may detect a certain stage in the keratinocyte terminal differentiation program but, to date, the chemical bases of the Ayoub–Shklar and rhodanile blue staining of keratinocytes has not been determined.

D. The Effect of Hydrocortisone on Keratinocyte Colony Morphology and Growth

In the first attempts to culture human foreskin keratinocytes with a 3T3 feeder layer, hydrocortisone was added to the medium because it inhibited stratification in the earlier studied mouse keratinocyte line XB (Rheinwald and Green, 1975a). It was therefore suspected that hydrocortisone might enhance proliferation at the expense of terminal differentiation in human keratinocytes. It was found instead that the steroid had no effect on colony morphology, growth rate, or colony-forming efficiency in primary human keratinocyte cultures. In secondary and subsequent cultures, however, keratinocytes do not assume normal epithelioid morphology during the early stage of colony initiation in the absence of hydrocortisone (Fig. 4), although colony-forming efficiency is unaffected. Cells growing in the absence of hydrocortisone eventually organize into a normal colony morphology by the time colonies have reached a size of about 100 cells. However, when 3T3 conditioned medium is substituted for intact 3T3 feeder cells to support keratinocyte growth (see below), human epidermal keratinocytes demonstrate an absolute requirement for hydrocortisone in order to initiate colonies from single cells (J. G. Rheinwald and H. Green, unpublished observations).

E. The Behavior of Keratinocyte Colonies after Selective Removal of the Feeder Layer

Keratinocyte colonies initiated in the presence of a 3T3 feeder layer respond to removal of the feeder cells by a rapid expansion of colony area. The peripheral cells of each colony spread out greatly to become very flat, and the more centrally situated cells follow suit to a degree related to their distance from the colony periphery. In small colonies (less than about 30 cells) all the cells become very flat and cease dividing. In large colonies (greater than about 150 cells), centrally located cells become only slightly more flattened after the feeder layer is removed, and exponential growth of such colonies continues until the culture is confluent (Fig. 5). Colonies of intermediate size (30–150 cells) have a probability of survival after feeder layer removal proportional to their size. Densely plated single-cell suspensions of keratinocytes artifactually display an elevated frequency of small colony survival after the feeder layer has been removed, because neighboring colonies often merge to form pseudocolonies sufficiently large to continue growth. Large colonies do not require and are not stimulated by conditioning of the medium by 3T3 cells or by the addition of hydrocortisone after the feeder layer has been removed, while the survival frequency of small and intermediate colonies is increased by these factors.

F. The Initiation of Keratinocyte Growth in the Absence of a Feeder Layer

Many previous investigators had overlooked the requirement for fibroblast support because a pure population of keratinocytes could grow and express differentiated functions when plated at high density in standard culture media. Successful high-density primary cultures of human [reviewed by Flaxman (1974) and Karasek (1975)], rabbit (Liu and Karasek, 1978), mouse (Yuspa *et al.*, 1970; Fusenig, 1971), and guinea pig (Regnier *et al.*, 1973) epidermal keratinocytes have been reported. The precipitous fall-off in growth that results when

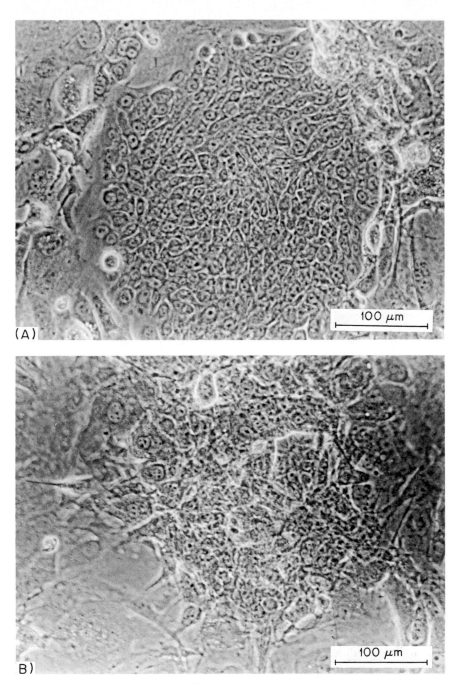

FIG. 4. Human epidermal keratinocyte colonies in a secondary culture growing with a 3T3 feeder layer in the presence of 0.4 μg/ml hydrocortisone (A) and the absence of hydrocortisone (B).

keratinocytes are plated in standard medium at densities less than 5×10^4 cells/cm² (Fig. 6) appears to be a result of the decreased probability, as plating density is reduced, that cells will be able to coalesce into pseudocolonies of sufficient size to grow under ordinary culture conditions.

Human keratinocytes are not stimulated to grow very much by 3T3 cells that share the culture fluid, but are not in direct contact with the keratinocytes. In the experiment summarized in Fig. 3 keratinocyte colony-forming frequency decreases dramatically as the 3T3 feeder layer density is reduced to less than 1.5×10^4/cm², even when medium conditioned by a fully confluent 3T3 culture is used as the growth medium. In another type of experiment, in which an optimal number of 3T3 feeder cells is permitted to attach to only half the surface area of a culture dish before a suspension of keratinocytes is added to the entire dish, colonies form only on the portion of the dish containing the feeder cells. This

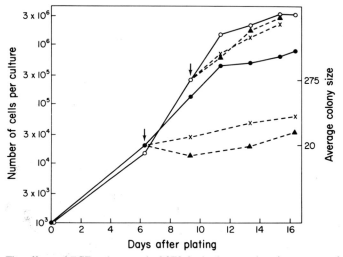

FIG. 5. The effects of EGF and removal of 3T3 feeder layer on keratinocyte growth. Foreskin strain E cells (10⁴ cells, 21 cell generations in culture at the time of plating) were plated with 2×10^4 γ-irradiated 3T3 cells/cm² and gave rise to 950 keratinocyte colonies per 60-mm dish. EGF was added to half the cultures on the fourth day after plating. Beginning on the sixth day after plating, the number of keratinocytes per culture was determined. On days 6 and 9, the feeder layer was removed from some dishes and the cultures were incubated further in regular medium in the presence (triangles) or absence (X) of EGF. Solid circles, Cultures grown with the 3T3 feeder layer minus EGF; open circles, cultures grown with the 3T3 feeder plus EGF. Cultures grew poorly when the feeder layer was removed at day 6, and the average colony size was 22, but grew even better than the feeder layer cultures when the feeder layer was removed at day 9 at a time when the average colony size was 275. The addition of EGF resulted in much better growth of colonies growing with the feeder layer, but had no effect on colony growth after the feeder layer was removed. (In this experiment, the feeder layer was removed from minus-EGF cultures on day 6 and from plus-EGF cultures on day 9, but the behavior of the colonies after removal of the feeder layer was the same whether prior culture was with or without EGF.)

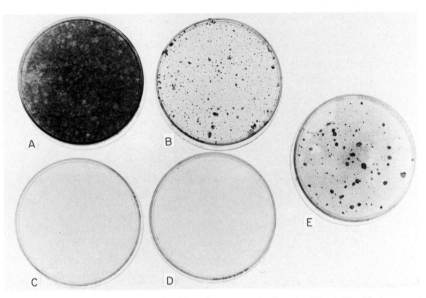

FIG. 6.   Density-dependent survival of keratinocytes subcultured without fibroblast support. Human keratinocyte strain N (27 cell generations in culture at the time of plating), previously passed with a 3T3 feeder layer and EGF, was subcultured at 10^6 (A), 3×10^5 (B), 10^5 (C), and 3×10^4 (D) cells per 60-mm dish in DME plus 20% FCS plus 0.4 μg/ml hydrocortisone without 3T3 feeder cells, or at 5×10^3 cells per 60-mm dish in the same medium with 3T3 feeder cells (E). All cultures were fixed 12 days later and stained with rhodanile blue.

requirement for contact with 3T3 cells is partially satisfied by plating the keratinocytes on a culture vessel surface that has previously supported a confluent monolayer of 3T3 cells (Fig. 7). When such a "3T3-conditioned dish" is used, 3T3-conditioned medium and hydrocortisone are also required and colony-forming frequency is typically less than 20% of that obtained when the keratinocytes are coplated with the optimal density of 3T3 feeder cells. The colony-stimulated substances left behind after the removal of 3T3 cells with EDTA can be removed by trypsin (J. G. Rheinwald and H. Green, unpublished observations). 3T3-conditioned medium 199 is superior to conditioned DME for promoting clonal growth in the absence of a 3T3 feeder layer, and raising the hydrocortisone concentration from 0.4 to 10 μg/ml increases the frequency of colony formation (D. M. Peehl and R. G. Ham, personal communication).

G.   The Expression of Tissue-Specific Structure and Function by Cultured Keratinocytes

Examination of living keratinocyte colonies by phase-contrast optics reveals a multilayered structure. The cell layers are sometimes thicker and more disor-

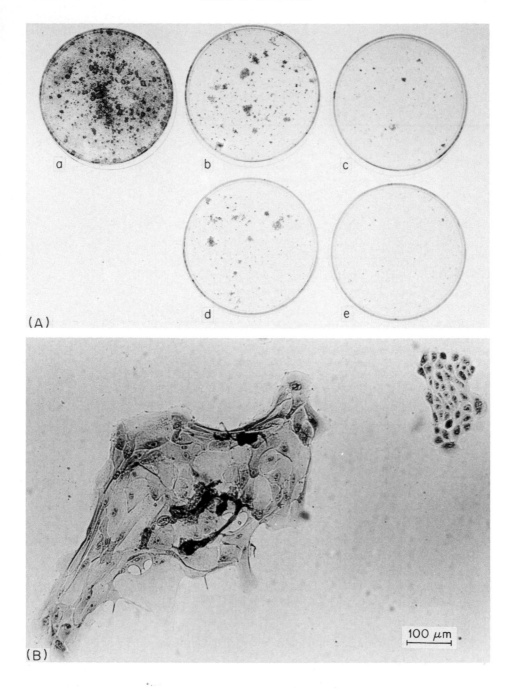

ganized at the point of colony initiation, which then usually remains as the colony's geographic center (Fig. 1). A flatter and more regularly organized layer of large, overlapping squamoid cells forms over most of the colony, leaving only the colony periphery as a single-cell layer. The dividing portion of a growing colony is restricted to the single layer of small, polygonal epithelioid cells attached directly to the culture vessel surface (Rheinwald and Green, 1975b), including both the cells at the colony periphery [whose [³H] thymidine (TdR) incorporation can easily be demonstrated by autoradiography of whole colonies] and those more centrally located beneath the squamous layers (whose nuclei are at a distance from an overlaid photographic emulsion greater than the range of the tritium β particle and must be seen by ¹⁴C[TdR] incorporation or by autoradiography of thin sections) (Rheinwald and Green, 1975b). *In vivo,* the basal layer of keratinocytes (the cells in direct contact with the underlying connective tissue via a basal lamina) is normally the sole dividing compartment of stratified squamous epithelia (Leblond *et al.,* 1964). Population pressure within the basal layer randomly forces some cells to lose contact with the connective tissue and move outward, where they cease dividing, embark upon an irreversible pathway of terminal differentiation, and are eventually shed from the free outer surface of the epithelium. That the upper layers of large, flattened squamous cells are derived from the dividing layer of small cuboidal cells in keratinocyte colonies growing *in vitro* was confirmed by pulse-chase experiments with ³H[TdR] (Sun and Green, 1976).

Throughout their culture lifetime, keratinocytes produce keratins as 30–40% of their total cell protein (Sun and Green, 1978a). This group of about six proteins has now definitely been demonstrated to be the epithelial cell-specific cytoplasmic 80-Å tonofilament proteins (Sun and Green, 1978b). *In vivo* and *in vitro,* when a terminally differentiating keratinocyte dies and its cytoplasm no longer maintains a reducing oxidation state, the keratin proteins become crosslinked by disulfide bonds and therefore require a reducing agent such as mercaptoethanol or DTT in addition to SDS or urea to become soluble (Sun and Green, 1978a).

The final product of keratinocyte terminal differentiation is the cornified envelope—an alkali-insoluble and ultrastructurally electron-dense layer that

---

Fig. 7. Initiation of colony growth from low-density platings of keratinocytes without intact 3T3 feeder cells. (A) Strain N keratinocytes (2×10^4 cells, 35 cell generations in culture at the time of plating) were plated in DME plus 20% FCS plus 0.4 μg/ml hydrocortisone with 3T3 feeder cells (a), or in 3T3-conditioned (M199 plus 20% FCS) plus 10 μg/ml hydrocortisone in a 3T3-conditioned dish (b), a regular dish (c), a 3T3-conditioned dish plus 20 ng/ml EGF (d), or a regular dish plus 20 ng/ml EGF (e). All cultures were fixed 12 days later and stained with rhodanile blue. (B) Two colonies growing in dish (b) above. Growth in the colony at the left was abortive, and all cells have become greatly flattened and enlarged. The colony at the right consists of small healthy cells and is actively growing.

forms beneath the plasma membrane (Matoltsy and Balsamo, 1955; Farbman, 1966). These structures are proteinaceous and can be observed under the light microscope with the aid of phase-contrast or Nomarski optics as transparent sacks remaining after whole epidermis or cultured keratinocytes have been extracted with SDS or urea plus a reducing agent (Sun and Green, 1976). The cornified envelope has now been characterized as the product of a cytoplasmic transglutaminase that catalyzes isodipeptide cross-linking of a particular keratinocyte cytoplasmic protein in terminally differentiating cells that have lost selective membrane permeability (Rice and Green, 1977, 1978).

H.   The Triggering of Keratinocyte Terminal Differentiation by Suspension

Although the dividing basal layer of growing keratinocyte colonies is continually giving rise to cells that move upward, flatten, enlarge, disulfide-link their keratins, and eventually form a cornified envelope, the majority of the cells in a growing colony are basal cells cycling with an intermitotic time of about 24 hours (Rheinwald and Green, 1975b). The proportion of cells that reinitiate growth after subculturing such colonies with the feeder layer system is 10% at best. That the number of these colony-forming cells is always smaller by a factor of 5 or more than the number of cells actively cycling in the colonies prior to subculture, suggests that the process of disaggregation and replating may be particularly harmful to the keratinocytes. *In vivo,* the keratinocytes of the basal layer cease dividing and terminally differentiate after they lose contact with the basal lamina and move outward. The cause-and-effect relationship between this loss of substratum contact and initiation of the terminal differentiation program is unknown, but in culture a brief absence of contact with a suitable attachment substratum is sufficient to trigger terminal differentiation in keratinocytes.

Normal human fibroblasts, while requiring anchorage and spreading in order to divide, enter a fully reversible nondividing state when held in suspension; keratinocytes lose their ability to reinitiate growth in a surface culture with a $t_{\frac{1}{2}}$ of about 3 hours (Rheinwald and Green, 1977; also Fig. 8). The probable length of time required for keratinocytes to reattach and spread after being dissociated and reinoculated accounts for most of the difference in the proportion of cells demonstrating division potential before and after subculture. It has since been demonstrated (Green, 1977) that the loss of colony-initiating ability by anchorage-deprived keratinocytes is followed by an ordered terminal differentiation sequence, triggered in previously dividing basal cells by detachment from the culture vessel surface. The next step is the loss of selective permeability by the plasma membrane and cross-linking of the keratin proteins. The cornified envelope then forms, and ultimately the nucleus and cytoplasmic organelles are digested to yield cells that are biochemically and ultrastructurally similar to the

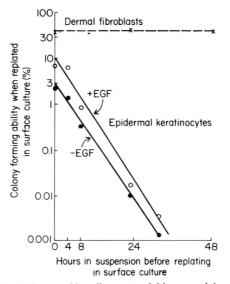

FıG. 8. Ability of diploid human skin cell types to reinitiate growth in surface cultures after being held in suspension in a nondividing state. The keratinocytes and fibroblasts were derived from the same newborn foreskin and were both at 30 cell generations in culture when placed in suspension. Suspension medium was 3T3-conditioned (DME plus 20% FCS) containing 1.3% methylcellulose. After incubating in suspension at 37°C for various time periods, the cells were plated in surface culture (in DME plus 20% FCS plus 0.4 μg/ml hydrocortisone with 2×10^4 3T3 feeder cells/cm^2 for keratinocyte clonal growth, or with DME plus 20% FCS with 2×10^3 3T3 feeder cells/cm^2 for fibroblast clonal growth). Keratinocytes were grown and subsequently incubated in suspension in the presence (open circles) or absence (solid circles) of 20 ng/ml EGF. (X) indicates fibroblasts grown and subsequently incubated in suspension in the absence of EGF.

mature stratum corneum cells of keratinizing epithelium *in vivo*. All these changes occur in cells growing in surface culture, but at a slower rate than in suspension and only in cells that have moved into the upper layers of growing colonies (Sun and Green, 1976).

I.   The Effect of Epidermal Growth Factor on Cultured Keratinocytes

Keratinocyte colonies growing with a 3T3 feeder layer respond to the addition of 1–30 ng/ml of EGF in the following way (Rheinwald and Green, 1977). Within 24 hours, colonies become more spread out and are less thickly stratified—much in the same way as colonies from which the surrounding feeder cells have been removed, but to a lesser degree. The cells maintain a two- to fivefold elevated colony-forming ability at subculture (relative to parallel cultures growing without EGF) and continue to grow exponentially until the culture becomes confluent (Fig. 5). Colonies growing in the absence of EGF, particularly in secondary and later passages, slow down from their maximum growth rate and assume a greatly decreased colony-forming ability long before neighboring colonies touch and they become spatially restricted from further expansion by the surface area of the culture dish. EGF is routinely added at each passage at the time of the first refeeding, because it slightly reduces the frequency of colony formation when added at the time of inoculation. A concentration of 20 ng/ml now appears to be optimal; higher concentrations cause excessive flattening of the colonies and a decreased growth rate.

Keratinocytes from newborn foreskin can undergo over 160 cell generations in culture if grown in the presence of EGF and subcultured at each passage while they are still in the exponential phase of growth (Rheinwald and Green, 1977). Colony-forming ability is $\geqq 2\%$ during most of this time. The cells remain diploid and retain their keratinocyte-specific markers and ability to terminally differentiate. While EGF appears to forestall terminal differentiation in growing keratinocyte colonies, it has no effect on the kinetics of terminal differentiation in suspended cells (Rheinwald and Green, 1977; also Fig. 8). Cells growing in the absence of a 3T3 feeder layer—either colonies initiated from single cells in a conditioned medium–conditioned dish system or established colonies continuing to grow after the feeder layer has been selectively removed—are not stimulated by EGF at any concentration and are inhibited by EGF at 20 ng/ml (Fig. 7A). This is possibly a result of the excessive flattening EGF induces in keratinocytes not surrounded by a 3T3 feeder layer.

IV.   Discussion

Two major departures from contemporary notions about cultured cells permitted successful cultivation of the human epidermal keratinocyte. The first was the use of a feeder layer of a heterologous cell type to achieve clonal growth. Feeder layers of the same cell line whose clonal growth is desired have been used for many years to supply nutrients not present in a suboptimal medium (Fisher and Puck, 1956). The functions provided specifically by a fibroblast feeder layer that permit clonal growth of keratinocytes are still not completely defined, but include substratum modification, secretion into the medium of a heat- and protease-labile macromolecule, and reduction of apparently inhibitory concentrations of cystine and tyrosine in the DME (J. G. Rheinwald and H. Green, unpublished observations).

The second departure was from a way of thinking about the mammalian cell cycle that has become rather solidified as a generality from the cultured fibroblast model. A cell type such as the keratinocyte might not enter a reversible G_0 state at confluence, but instead might terminally differentiate when its growth rate slows. In recognition of this, in order to determine the *in vitro* growth requirements and replicative capacity of the keratinocyte, cultures have been initiated and passaged at low densities, and colonies subcultured while they are still small. Confluent cultures contain a very low proportion of cells able to reinitiate growth after subculture, which explains the inability of previous investigators to passage keratinocytes serially without a feeder layer even at high density.

The prospect that the fibroblast feeder layer interacts with the keratinocytes physiologically, if not spatially, as do the fibroblasts of the supporting dermis *in vivo,* makes the *in vitro* response of keratinocytes to EGF and hydrocortisone rather interesting. EGF exerts a large stimulatory effect on keratinocytes growing with a feeder layer, but not without. In fact, removal of the feeder layer from well-established keratinocyte colonies seems to mimic the EGF effect, possibly by changing the shape of the cells or by abolishing an inhibition of colony expansion imposed by the feeder cells. It has been found that corneal keratinocytes [which display the same response to EGF in the 3T3 feeder system as described above for epidermal keratinocytes (Sun and Green, 1977)] fail to respond to EGF when grown directly on plastic without a feeder layer but are stimulated by EGF when grown on a collagen gel (Gospodarowicz *et al.,* 1978). Hydrocortisone effects packing into an organized epithelioid arrangement by colonies initiating in the feeder layer system and is an essential growth factor for cells trying to initiate clonal growth in the absence of intact feeder cells. The effects of EGF and hydrocortisone and the requirement for factors produced by fibroblasts are obscured after keratinocytes have grown into large colonies or when they are plated at very high density. Normally, adult stratified squamous epithelia *in vivo* contain a tightly packed, high-density, dividing basal cell population, so it may be that these factors play their most important regulatory roles during embryogenesis and in wound healing.

The intrinsic tendency of the keratinocyte to stratify and terminally differentiate even in the absence of instructive information from the connective tissue stroma was first described by Flaxman *et al.* (1967) in their study of primary explant outgrowth cultures of epidermis. These workers noticed that keratohyalin granules—the irregularly shaped electron-opaque inclusions that appear in the layer of cells just beneath the stratum corneum of the epidermis—were not present in the stratified keratinocyte outgrowths. Keratohyalin granules have not been observed in clonal cultures with a feeder layer system or during terminal differentiation of suspended cells. Nonetheless, the most important features of keratinocyte differentiation—disulfide-linking of keratins, cornified envelope formation, and enucleation—occur in culture to yield a stratum corneum-like cell without the requirement for a granular cell intermediate. The highest-molecular-weight keratin (~63K) found in stratum corneum *in vivo* is not synthesized by cultured keratinocytes, however (Sun and Green, 1978a), and the staining characteristics of the keratin filament matrix in the end-stage cells that have differentiated *in vitro* (Green, 1977) is somewhat different from that in stratum corneum cells *in vivo* (Brody, 1960).

The details of basal cell shape and the number of intermediate and squamous layers are not precisely reproduced in the tissue culture plastic and fibroblast feeder system (Rheinwald and Green, 1975b). Clearly, to permit the study of some of the more subtle features of keratinocyte growth control, tissue organization, and ultrastructural change during terminal differentiation, a better *in vitro* reproduction of the natural environment is required. Perhaps the most important component of the culture system to improve is the substratum. Epithelial tissues are fed from beneath the basal layer by diffusion from the vascularized dermis, rather than from above via diffusion through the squamous cell layers to the dividing basal cells as are colonies growing *in vitro*. Offering a more natural substratum and improved nourishment of the dividing basal cell layer may create a culture environment permissive for the complete reproduction of tissue structure. Studies of intrinsic differentiative potential versus instructive influences of embryonic mesenchyme and adult connective tissue would then be feasible.

The apparent replicative life span of human keratinocytes in culture is subject to wide fluctuation depending upon whether terminal differentiation is encouraged (e.g., by maintaining cells in a slowly dividing or nondividing state) or discouraged (e.g., by growth in the presence of EGF and subculturing when colonies are relatively small). Senescence eventually occurs even when the currently optimal passage regimen is used, so it appears that a more general metabolic system in keratinocytes may deteriorate during replication *in vitro,* as has been observed for fibroblasts. The relative importance of the tissue-specific terminal differentiation program and a possibly more general mechanism of replicative senescence in the determination of keratinocyte culture lifespan is as yet undetermined. The human epidermal basal cell divides on the average of once every 2 weeks *in vivo* (Weinstein and Frost, 1969), suggesting that keratinocytes have an inherent replicative capacity of several thousand divisions. Further attempts to improve the culture system might be worthwhile to attain more nearly this doubling potential.

V.   Perspectives

The keratinocyte culture system described here is more complex than standard methods only in the use of the easily cultivated 3T3 cell line as a feeder and in the necessity of subculturing the keratinocyte population before it leaves the exponential growth phase. The newborn human foreskin is a readily available source of both keratinocytes and fibroblasts with high proliferative capacity, permitting *in vitro* experiments with parallel populations of very different cell types having identical genotypes.

Novel mechanisms of growth factor action have been discovered in cultured keratinocytes. EGF appears to stimulate keratinocyte growth by altering colony morphology and reducing the rate at which hitherto dividing cells enter the terminal differentiation program (Rheinwald and Green, 1977). Recently Green (1978) has reported that agents that increase cellular cyclic AMP content can increase the growth rate of cultured keratinocytes. Other key elements controlling progress through the cell cycle are likely to be different in the keratinocyte and other cell types from those in the well-studied fibroblast, and merit study as possible points at which lesions occur in neoplasia.

A major problem in studying carcinogenesis and neoplastic development in human cells is that the naturally arising fibrosarcoma of the easily cultivated dermal fibroblast is very rare and not easily cultivated. As a result, the characteristics of naturally occurring malignant variants of the fibroblast are unknown, and experimentally virus-transformed fibroblast lines offer the only information on what attributes should be looked for. On the other hand, squamous cell carcinomas of the epidermis and the oral epithelia are very common. The malignant cells can readily be identified in primary cultures of tumor biopsies and purified of associated normal colony-forming keratinocytes (Rheinwald and Beckett, 1980). Squamous carcinoma cells behave very differently in a number of respects from normal keratinocytes but do not mimic the normal versus virus- or chemical-transformed rodent fibroblast model that has been well-characterized in culture.

The ability to identify malignant epithelial colonies in primary cultures of tissue biopsies is potentially applicable as a tool for diagnosing possible neoplastic lesions in keratinizing epithelia. *In vitro* carcinogenesis studies of human keratinocytes, where the appearance of colonies possessing carcinoma-specific phenotypes is followed, promises to yield important information about the tissue-specific potency of primary carcinogens and promoters.

The keratinocyte expresses portions of the human genome that the fibroblast does not. Human wart virus replication (which is not supported by fibroblasts) and its role in producing benign neoplasia is potentially amenable to study in keratinocyte cultures. In addition to the well-studied keratin–tonofilament proteins (Sun and Green, 1978a; Fuchs and Green, 1978) produced in abundance by this cell in culture, enzymes and cell surface proteins specific to keratinocytes or to epithelial cells in general can now be readily studied and the biochemical bases of heritable diseases whose genetic defects are not expressed by fibroblasts may be determined. Chromosome assignments of many new genes should be possible using interspecific hybrids with cultured human keratinocytes.

Acknowledgments

It is a pleasure to acknowledge the support, advice, and encouragement of Howard Green, in whose laboratory most of the original research reviewed here was completed.

References

Ayoub, P., and Shklar, G. (1963). *J. Oral Surg.* **17,** 580–581.
Brody, I. (1960). *J. Ultrastruct. Res.* **4,** 264–297.
Farbman, A. I. (1966). *Anat. Rec.* **156,** 269–282.
Fisher, H. W., and Puck, T. T. (1956). *Proc. Natl. Acad. Sci. U.S.A.* **42,** 900–906.
Flaxman, B. A. (1974). *In Vitro* **10,** 112–118.
Flaxman, B. A., Lutzner, M. A., and Van Scott, E. J. (1967). *J. Invest. Dermatol.* **49,** 322–332.
Fuchs, E., and Green, H. (1978). *Cell* **15,** 887–897.
Fusenig, N. E. (1971). *Naturwissenchaften* **8,** 421.
Goldberg, B. (1977). *Cell* **11,** 169–172.
Gospodarowicz, D., Greenburg, G., and Birdwell, C. R. (1978). *Cancer Res.* **38,** 4155–4171.
Green, H. (1977). *Cell* **11,** 405–411.
Green, H. (1978). *Cell* **15,** 801–811.
Jacobs, L., and DeMars, R. (1977). *In* "Handbook of Mutagenicity Test Procedures" (B. Kilbey *et al.,* eds.), pp. 193–220. Elsevier, Amsterdam.
Karasek, M. (1975). *J. Invest. Dermatol.* **65,** 60–66.
Leblond, C. P., Greulich, R. C., and Pereira, J. P. M. (1964). *Adv. Biol. Skin* **5,** 39–67.
Liu, S. C., and Karasek, M. (1978). *J. Invest. Dermatol.* **70,** 288–293.
MacConaill, M. A., and Gurr, E. (1964). *Ir. J. Med. Sci.* [7] 243–250.
Matoltsy, A. G., and Balsamo, C. A. (1955) *J. Biophys. Biochem. Cytol.* **1,** 339–360.
Pious, D. A., Hamburger, R. N., and Mills, S. E. (1964). *Exp. Cell Res.* **33,** 495.
Regnier, M., Delescluse, C., and Prunieras, M. (1973). *Acta Derma.-Venereol.* **53,** 241.
Rheinwald, J. G., and Beckett, M. A. (1980). In preparation.
Rheinwald, J. G., and Green, H. (1975a). *Cell* **6,** 317–330.
Rheinwald, J. G., and Green, H. (1975b). *Cell* **6,** 331–344.
Rheinwald, J. G., and Green, H. (1977). *Nature (London)* **265,** 421–424.
Rice, R. H., and Green, H. (1977). *Cell* **11,** 417–422.
Rice, R. H., and Green, H. (1978). *J. Cell Biol.* **76,** 705–711.
Rothfels, K. H., and Siminovitch, L. (1958). *Stain Technol.* **33,** 73.
Savage, C. R., and Cohen, S. (1972). *J. Biol. Chem.* **247,** 7609–7611.
Steinberg, M. L., and Defendi, V. (1979). *Proc. Natl. Acad. Sci. U.S.A.* **76,** 801–805.
Sun, T.-T., and Green, H. (1976). *Cell* **9,** 511–521.
Sun, T.-T., and Green, H. (1977). *Nature (London)* **269,** 489–493.
Sun, T.-T., and Green, H. (1978a). *J. Biol. Chem.* **253,** 2053–2060.
Sun, T.-T., and Green, H. (1978b). *Cell* **14,** 469–476.
Tjio, J. H., and Puck, T. T. (1958). *J. Exp. Med.* **108,** 259.
Todaro, G. J., and Green, H. (1963). *J. Cell Biol.* **17,** 299–313.
Weinstein, G. D., and Frost, P. (1969). *Natl. Cancer Inst., Monogr.* **30,** 225–246.
Yuspa, S. H., Morgan, D. L., Walker, R. J., and Bates, R. R. (1970). *J. Invest. Dermatol.* **55,** 379–389.

Chapter 16

Dermal Fibroblasts[1]

RICHARD G. HAM

Department of Molecular, Cellular and Developmental Biology,
University of Colorado,
Boulder, Colorado

[1]The research described in this chapter was supported by grant CA 15305 from the National Cancer Institute, grant AG 00310 from the National Institute on Aging, and contract 223-74-1156 from the Bureau of Biologics, U.S. Food and Drug Administration.

I.   Introduction

For most of the types of cells described in this volume, the major problem is how to obtain satisfactory growth in culture. This problem does not have to be faced with fibroblasts. In fact, it is often necessary to deal with the opposite problem—how to prevent cultures of other types from being totally overrun by fibroblasts, which some investigators undoubtedly view as "weeds" in the garden of exotic cell types.

Since detailed procedures for initiating and maintaining fibroblast cultures have already been well described in the literature and in standard technique manuals, only a brief summary of standard procedures is presented in this chapter. Instead, our emphasis is on recent progress that has been made toward understanding the complete set of requirements for the multiplication of human diploid fibroblasts and on new media and techniques that have been developed for the growth of fibroblasts with minimal amounts of serum protein.

Differences in growth requirements between fibroblasts from newborn foreskin and from fetal lung are also discussed, as well as differences between the growth requirements of fibroblasts and those of various types of epithelial cells. The latter differences suggest the interesting possibility of developing culture media that strongly favor multiplication of a particular type of cell. If this can be accomplished, it may in the future be possible through use of appropriate media to select either for or against the growth of fibroblasts in primary cultures. Preliminary comparisons of the growth requirements of dermal fibroblasts and keratinocytes, which are described later in this chapter, appear to indicate that such a selection is feasible.

II.   Standard Procedures for Dermal Fibroblasts

A.   Media and Sources of Tissue

Dermal fibroblasts are widely used as research tools in many fields, including human genetics and genetic counseling. Although minor variations exist among the procedures recommended by various authors, these procedures are basically so similar that they can be viewed as standard.

When adequate amounts of serum are used, almost any standard cell culture medium will support satisfactory growth of fibroblasts, either of dermal or lung origin. Frequently used media include Eagle's basal (BME), Eagle's minimum essential (MEM), Dulbecco's modification of Eagle's basal (DME), McCoy's 5a, F12, and others. Fetal bovine serum is usually used at concentrations between 10 and 20% (v/v), although calf serum has also been reported to be satisfactory. Toxicity testing and selection of serum lots is desirable when high concentrations of serum are used.

The usual sources of tissue are punch biopsies and foreskins. Punch biopsies provide rather limited amounts of tissue but allow the establishment of cultures from individuals of any age and of both sexes who have been preselected for particular traits of interest to the investigator establishing the cultures. Foreskins provide much larger tissue samples but are available only from newborn male infants with little or no genetic selection.

Cultures are generally initiated either as outgrowths from tissue fragments or by inoculating with cell suspensions prepared by enzymatic digestion of the tissue. Outgrowth from fragments is strongly preferred for the small tissue samples obtained from punch biopsies and is also frequently used for foreskin cultures. The use of primary cell suspensions has the advantage of providing larger cultures more rapidly but is complicated by problems of dissociating the tightly adherent skin cells to form a viable cell suspension.

B.   Outgrowth from Tissue Fragments

Highly successful techniques for the establishment of fibroblast cultures from tissue fragments have been described repeatedly in the literature (e.g., Greene, 1973; Martin, 1973; Goetz, 1975; Simpson and Stulberg, 1975; Schneider and Mitsui, 1976). Most investigators prefer punch biopsies between 2 and 8 mm in diameter, and a depth of 2–3 mm or to the subcutaneous fat. Martin (1973) claims that good results can routinely be obtained with diameters as small as 4 mm and depths as little as 1 mm. He also reports successful establishment of vigorous cultures from "pinch" biopsies less than 1 mm in diameter.

Most authors recommend use of a sharp scalpel to cut tissue samples (either punch biopsy or foreskin) into fragments approximately 1 mm across and free of subcutaneous fat. Tearing the tissue or cutting with scissors is not recommended. The most important single requirement for initiating outgrowth cultures is to maintain tight contact between the fragments and the culture surface until the outgrowth is established. This is usually accomplished either by keeping the volume of medium so small that the fragments cannot float or else by using a weight (such as a microscope slide) to hold the fragments against the culture surface. Some investigators use a very small volume of medium and then invert the culture vessel, such that the fragment is held tightly against the culture surface by the adherent film of medium.

When a small volume of medium is used to keep the tissue fragment in contact with the culture surface, adequate humidification is important to prevent evaporation or increase in osmolarity of the culture medium. As soon as the fragment has attached firmly to the culture surface, the volume of medium should be increased so there is no depletion of nutrients or buildup of waste products in the microenvironment surrounding the fragment.

Epithelial outgrowth usually appears first around a tissue fragment but is soon overrun by a more migratory population of elongated fibroblast-like cells. Because of the natural tendency of the fibroblastic cells to dominate primary cultures, it is generally not necessary to take special steps to eliminate other types of cells present in the initial inoculum.

C.   Primary Cell Suspensions

The major problems with the use of cell suspensions to establish primary cultures are the difficulty experienced in disaggregating the tissue and the relatively low viability of the single cells obtained. Pious et al. (1964) used 0.05% (w/v) trypsin in conjunction with 6×10^{-4} M EDTA to obtain a suspension that yielded both epithelial and fibroblastic colonies in medium 199 supplemented with 15% fetal calf serum. Howard et al. (1976) used a 24-hour digestion with 0.25% (w/v) trypsin prepared in Eagle's MEM, followed by filtration through lens paper to remove undigested fragments. The cell suspension was centrifuged at 500 g for 10 minutes, and the resultant cell pellet was dispersed in medium F12 containing 20% fetal bovine serum. The method was reported to yield a culture of approximately 10^7 cells per foreskin within 2 weeks. Heckman et al. (1977) prepared a cell suspension from minced foreskin by incubation with occasional stirring in 0.2% (w/v) collagenase and initiated fibroblast cultures in Eagle's MEM supplemented with 10% fetal calf serum.

Rheinwald and Green (1975) used sequential 30-minute digestions with 0.25% (w/v) trypsin to prepare a cellular suspension from minced foreskin tissue. Their objective was growth of keratinocytes, but the inoculum was also rich in fibroblasts, and it was necessary to use a feeder layer to prevent the keratinocyte cultures from being overrun with fibroblasts in Dulbecco's modified Eagle's medium supplemented with 20% fetal calf serum and 0.4 μg/ml hydrocortisone. In our laboratory, we use 0.25% (w/v) trypsin plus 6.8×10^{-4} M EDTA to prepare a cell suspension from minced foreskin that can be used either for the growth of keratinocytes or fibroblasts, depending on the culture medium employed, as will be discussed in Section VI, B.

D.   Serial Culture

Serial cultivation of fibroblast cultures is usually done by dispersing the cells with a relatively dilute (0.05% w/v) trypsin solution, although Schneider and Mitsui (1976) prefer to use a minimal volume of 0.1% (w/v) pronase.

The karyotype and other normal properties are generally quite stable in human skin fibroblast cultures. Aneuploidy is relatively rare, even in old cultures, and transformation due to causes other than oncogenic viruses is essentially nonexistent. Thus, in the absence of deliberate viral transformation, cultures normally remain diploid and undergo a well-defined *in vitro* senescence.

The total proliferative capacity of skin fibroblast cultures varies greatly from one individual donor to another, but there is a distinct tendency toward shorter culture life spans from older donors (Martin *et al.*, 1970; Schneider and Mitsui, 1976). A progressive loss of proliferative capacity of individual cells occurs throughout the *in vitro* lifespan of the cultures (Smith *et al.*, 1978). Because of this, it is desirable to use cultures at relatively low population doubling levels whenever possible, particularly in clonal growth experiments.

III.   Procedures for Growth of Fibroblasts with Minimal Amounts of Serum Protein

A.   Holistic Approach to Culture Environment

The major emphasis of current research in our laboratory is precise definition of the requirements for multiplication in culture of normal diploid cells and the development of practical laboratory procedures that will permit normal cells to be grown with the least possible amount of undefined supplementation in the culture system. In these studies, we have found it necessary to take a holistic approach toward the total culture environment, in which every variable is considered a single member of a highly complex set so constituted that modification of any member of the set may affect the response to any other member of the set. The parameters that affect the amount of serum protein needed for growth have been found to go far beyond the composition of the culture medium and physiological variables such as temperature, pH, and osmolarity. In particular, we have found that use of a mild, low-temperature trypsinization technique to disperse the cells and the use of modified culture surfaces are both very important for the growth of human diploid fibroblasts with minimal amounts of serum protein, even when the most advanced media are employed.

The growth requirements of human skin and lung fibroblasts are quite similar (although not strictly identical, as discussed in Section V). These two cell types have been used almost interchangeably in the studies reported here. In many cases the observations were first made with lung fibroblasts (e.g., WI-38, MRC-5, IMR-90, or FLOW 2000) and subsequently verified with foreskin fibroblasts. Many of these observations have already been published and discussed in detail with emphasis on lung fibroblasts (McKeehan *et al.*, 1976, 1977, 1978;

McKeehan and Ham, 1976, 1978a,b; Ham *et al.*, 1977; McKeehan, 1977; Ham and McKeehan, 1978a,b, 1979.

In the sections that follow, particular emphasis is given to requirements for the growth of skin fibroblasts and to differences that have been observed between skin and lung fibroblasts.

B.   Optimized Medium

The development of improved media for human diploid fibroblasts has been described in detail elsewhere (McKeehan *et al.*, 1976, 1977; Ham *et al.*, 1977; Ham and McKeehan, 1978a,b, 1979). In summary, a sequential process of qualitative and quantitative optimization of medium composition for clonal growth with minimal amounts of serum protein resulted in the development of medium MCDB 104 (McKeehan *et al.*, 1977) and medium MCDB 105 (McKeehan *et al.*, 1978; McKeehan and Ham, 1978a). These two media are similar except that MCDB 104 is designed for use with 5% CO_2, whereas MCDB 105 is designed for use with 2% CO_2. Cellular multiplication with small amounts of serum protein is approximately equivalent in either medium. The principal advantage of MCDB 105 is that the use of less CO_2 in the gaseous phase minimizes the pH shift medium undergoes when removed from the incubator and thus reduces the amount of buffering needed. We now routinely use MCDB 105 with 2% CO_2 for monolayer and clonal growth of human diploid fibroblasts of skin and lung origin.

C.   Dialyzed Serum

One of the first steps taken in the development of an optimized medium for any type of cell in our laboratory is to adjust the culture conditions such that we can obtain satisfactory clonal growth with dialyzed serum as the primary undefined supplement. The reason for doing this is so that we can adjust the concentrations of all components of the synthetic medium to precisely determined quantitative optima with no interfering effects from small molecules in the serum. The basic procedure for the preparation of dialyzed serum is extended dialysis against several changes of distilled water, followed by centrifugation to remove the small precipitate that forms. The dialyzed serum is then lyophilized to a dry powder which is added to the culture medium in weighted amounts, usually from a 50 mg/ml stock solution in saline (McKeehan *et al.*, 1976). The lyophilized preparation is usually referred to as fetal bovine serum protein (FBSP), but it is important to keep in mind that it also contains substantial amounts of carbohydrates and lipids, as well as residual amounts of tightly bound low-molecular-weight nutrients such as fatty acids and trace elements.

D. Low-Temperature Trypsinization

A critical requirement for cellular multiplication with minimal amounts of serum protein is that the cells not be damaged during subcultivation. This has been achieved through the use of a low-temperature trypsinization procedure that employs the smallest possible volume of a trypsin solution of the lowest possible concentration for the shortest possible time that will adequately loosen the cells in question (McKeehan, 1977). For human foreskin fibroblasts previously grown to confluency in medium MCDB 104 (or 105) with 1000 μg/ml FBSP as the only supplement, the following procedure can be used.

Prepare in advance solution I (glucose, 4.0 mM; KCl, 3.0 mM; NaCl, 122 mM; Na$_2$HPO$_4$, 1.0 mM; phenol red, 3.3 μM; HEPES, 30 mM; adjusted to a final pH of 7.6 with NaOH). Save some of the solution I for rinsing the cells, and prepare 2$\times$ crystalline trypsin at a concentration of 500 μg/ml in the rest. The trypsin solution should be stored frozen until ready to be used and should have an activity of about 18 U/ml when measured under actual trypsinization conditions at 4°C in solution I.

The instructions that follow are for a 25-cm^2 flask. For larger flasks, increase all volumes proportionally to the surface area. Prior to subculturing, chill all solutions and media to be used in an ice bath and also chill the culture flask containing the cells and the original growth medium on ice for approximately 15 minutes before proceeding with the trypsinization. Remove the medium and rinse the monolayer twice with 0.5-ml aliquots of chilled solution I. Cover the cells with 2.0 ml of the trypsin solution, rock the flask gently to be certain that all cells have contacted the trypsin, and immediately withdraw as much of the solution as possible. Close the flask tightly to prevent evaporation and leave it in contact with an ice bath for about 15 minutes with only a thin film of trypsin solution over the cells. An inverted microscope may be used to check the progress of the trypsinization, but the flask should never be allowed to warm appreciably above ice bath temperature.

When the cells are partially loosened, add 5 ml of chilled serum-free medium and shake the flask gently to suspend the cells. Pipet gently as needed to break up clumps and dilute the cells as needed for monolayer or clonal subculture. If a hemocytometer count is done, keep the cell suspension on ice until it is completed. Do not warm the cells above the ice bath temperature until they are in their final culture medium and ready to be placed in the cell culture incubator.

It is necessary to determine the optimal trypsin concentration and trypsinization time empirically for the type of cell and the previous culture conditions. Fetal lung fibroblasts previously cultured under comparable conditions require only 100 μg/ml crystalline trypsin and approximately 5 minutes of digestion time for equivalent loosening [cf. McKeehan (1977) and Ham and McKeehan (1979) for details].

The trypsinization procedure described above significantly reduces the amount of serum protein needed for clonal growth of human diploid fibroblasts and increases both plating efficiency and colony size in clonal growth assays when FBSP is made rate-limiting for cellular multiplication. The lowered temperature causes a phase shift in the cellular membrane, which is thought to protect against mechanical damage and also to prevent pinocytosis of trypsin.

E.   Polylysine-Coated Culture Surfaces

The standard plastic dishes widely used for cell culture have a negative surface charge. We have found that human diploid fibroblasts multiply much better with minimal amounts of serum protein when the surface of the culture vessel is coated with a positively charged polymer (McKeehan and Ham, 1976). Any polymer with a positive charge appears to be effective. We have selected poly-D-lysine for routine use in our laboratory in order to minimize possible interference with other studies of growth-promoting substances. The following procedure can be used to coat petri dishes (or culture flasks):

Prepare a solution of 0.10 mg/ml poly-D-lysine–HBr (MW 30,000–70,000) in tissue culture-grade distilled water and sterilize it with a Millipore-type GS filter membrane (0.22 μm pore size). Add 0.5 ml to each 60 × 15 mm plastic tissue culture petri dish (or 25-cm² flask) and tilt gently to spread the solution uniformly over the entire culture surface. Allow the solution to stand in the dish for 5 minutes and then remove it as completely as possible by aspiration through a Pasteur pipet. Rinse the culture surface with 1.5 ml of sterile distilled water. The rinse solution must be removed completely, since free polylysine in the culture medium tends to inhibit cellular multiplication. The dishes can be filled with culture medium and used immediately, or they can be dried for later use. The entire coating procedure is usually done aseptically. However, it can also be done without sterile precautions and the coated dishes can be sterilized later with ultraviolet irradiation.

F.   Routine Culture Procedures

Through the use of medium MCDB 105 in conjunction with low-temperature trypsinization and polylysine-coated culture vessels, as described above, we obtain clonal growth of human diploid fibroblasts (both lung and skin) with as little as 10–25 μg/ml FBSP, and optimum growth occurs at 1000 μg/ml FBSP (equivalent in protein concentration to about 2.5% whole serum) as the only supplement. Monolayer growth is also excellent as long as the cultures do not become crowded, and we routinely grow serial cultures with 1000 μg/ml FBSP as the only supplement (generally without polylysine coating, which is of greater importance at lower protein concentrations). The doubling potential of such cultures is essentially unchanged, and they reach phase III at approximately the same number of population doublings as conventionally grown controls. The only limitation of the low-protein monolayer cultures is that density-dependent inhibition of multiplication occurs at a lower population density and crowding must be avoided to maintain a rapid multiplication rate.

IV.   Specific Requirements for Survival and Multiplication of Fibroblasts

A.   Research Strategy

The development of improved culture media and techniques for human diploid fibroblasts described above is part of a larger overall program of study of the growth requirements of a variety of types of normal cells both from humans and from other species. The research strategy employed in these studies has been discussed in detail elsewhere (Ham and McKeehan, 1978a,b, 1979) and will only be summarized briefly here.

In order to understand growth requirements fully, it is necessary to place cells in an environment that can be controlled precisely. For this reason, we have employed the clonal growth assay (Ham, 1972, 1974) in nearly all our studies. In this assay, the volume of medium per cell is so large that the individual cells cannot appreciably alter or "condition" their environment during the assay period. In addition, the small inoculum minimizes the carryover of growth factors from previous cultures, and any carryover that occurs either extracellularly or intracellularly is quickly depleted by the multiple rounds of replication each cell must undergo to form a colony.

In abbreviated form, the basic sequence of events in developing improved media is as follows [cf. Ham and McKeehan (1978a,b, 1979) for more detailed discussions]:

1. Employ whatever procedures that may be necessary to obtain cellular growth (not a problem with fibroblasts).

2. Adjust the culture medium and environment as needed to obtain clonal growth with dialyzed serum or other dialyzed supplements.

3. Reduce the concentration of the dialyzed supplement until it becomes rate limiting for colony formation and then attempt to restore optimal growth by making qualitative or quantitative changes in the synthetic portion of the medium.

4. Each time growth is improved, reduce the concentration of the dialyzed supplement until it is again growth limiting and attempt to achieve an additional improvement in clonal growth by further manipulation of the synthetic medium.

The optimized medium for human diploid fibroblasts, MCDB 105, contains all 20 of the amino acids involved in protein synthesis, 9 vitamins (biotin, folinate, lipoate, niacinamide, pantothenate, pyridoxine, riboflavin, thiamine, and vitamin B_{12}), 8 other organic components (adenine, choline, glucose, inositol, linoleate, putrescine, pyruvate, and thymidine), 7 major ions (calcium, magnesium, potassium, sodium, chloride, phosphate, and suffate), and 10 trace elements (copper, iron, manganese, molybdenum, nickel, selenium, silicon, tin, vanadium, and zinc). It also contains HEPES buffer and phenol red as a pH indicator. No bicarbonate is included in the basic formula, but a substantial amount of bicarbonate is formed when the buffered medium equilibrates with 2% CO_2 in the cell culture incubator.

Although qualitative changes such as the addition of trace elements and the substitution of adenine and folinic acid in place of hypoxanthine and folic acid are important, many of the improvements that distinguish MCDB 105 from medium F12 (Ham, 1965), which was the starting point for the development of MCDB 105, are quantitative. Our experience with growth requirements of various kinds of normal cells, both for different cell types from within the same species and for the same type of cell from different species, has shown the correct amount of a nutrient can be every bit as important as its qualitative presence. For example, most of the differences between the growth requirements of human dermal fibroblasts and human epidermal keratinocytes, which are discussed in Section VI are quantitative rather than qualitative. Such differences tend to be masked by large amounts of serum but are extremely important when the concentration of serum protein is reduced to the minimum that will support multiplication.

D.   The Requirement for Serum Protein

Under the survival conditions described in Section IV, B, human dermal fibroblasts require only a small amount of extensively dialyzed fetal bovine serum protein for rapid multiplication, either at clonal or monolayer density. Optimum clonal growth occurs with 500–1000 μg/ml FBSP, and reasonably good multiplication can normally be obtained with as little as 10–25 μg/ml.

The nature of the growth-promoting factors in the macromolecular fraction of serum is not yet fully understood. Although for convenience we frequently refer to the fraction as fetal bovine serum "protein," it also contains a substantial amount of lipid. The removal of lipids (Cham and Knowles, 1976) renders FBSP unable to support satisfactory multiplication of human diploid fibroblasts in culture. The addition of lecithin or synthetic phosphatidylcholine preparations to the delipidated serum protein in the form of water-soluble 250-Å-diameter phospholipid vesicles (liposomes) prepared by ultrasonic irradiation of phospholipid suspensions (Poste et al., 1976) restores multiplication-promoting activity for human diploid fibroblasts (McKeehan and Ham, 1978b). Natural soybean lecithin is highly active, but synthetic phosphatidylcholines are active only if they contain an unsaturated fatty acid (e.g., linoleic or oleic). Those containing only fully saturated fatty acids do not support multiplication of human fibroblasts.

Chicken embryo fibroblasts can be grown with delipidated serum supplemented with pure linoleic acid (Ham and McKeehan, 1978b), but we have been unable to obtain restoration of growth of human diploid fibroblasts in delipidated serum with pure fatty acids of any type. Data are still relatively incomplete, but lipids appear to be one of a small group of nutrients whose requirement for multiplication increases sharply as the concentration of (delipidated) fetal bovine serum protein is reduced (W. McKeehan, personal communication). Calcium ion behaves similarly (McKeehan and Ham, 1978a).

The specific protein factors needed for multiplication of human diploid fibroblasts are not yet fully characterized. We have, however, been able to achieve approximately a 15-fold purification with close to 50% recovery of biological activity (McKeehan et al., 1978). The fractionation scheme, which employs a combination of $(NH_4)_2 SO_4$ precipitation, polyethylene glycol precipitation, and pH precipitation, appears to separate the growth-promoting activity completely from the two major proteins of fetal bovine serum, fetuin and serum albumin. The active fraction exhibits multiple bands on sodium dodecyl sulfate (SDS) acrylamide gel electrophoresis and tends to lose activity during subsequent fractionation attempts, suggesting that it may still contain more than one species of biologically active material.

We have tested a number of the macromolecular growth factors described in the literature for various types of cells (Gospodarowicz and Moran, 1976). None of these will fully replace fetal bovine serum for clonal growth of human diploid dermal fibroblasts. Epidermal growth factor (EGF) has been reported to stimulate quiescent human foreskin fibroblasts (Cohen et al., 1975). We find that EGF significantly improves growth of foreskin fibroblasts in the presence of limiting amounts of fetal bovine serum protein, but that it is inactive when added alone to MCDB 105. Growth rate kinetic analysis, based on Michaelis–Menten enzyme rate kinetic analysis (McKeehan and Ham, 1978a), reveals that EGF does not change the amount of fetal bovine serum protein needed for half-maximum growth rate (analogous to K_m in Michaelis–Menten kinetics) and that its only effect is on the extrapolated "maximum growth rate" (analogous to V_{max} in Michaelis–Menten kinetics). Insulin and fibroblast growth factor have similar but less pronounced effects. The response to EGF appears to occur only with skin fibroblasts. Under identical experimental conditions, we are unable to detect any significant effect of EGF on clonal growth of human embryonic lung fibroblasts.

Conditioned medium from Buffalo rat liver cell cultures contains a multiplication-stimulating activity (MSA) for chicken embryo fibroblasts (Dulak and Temin, 1973). The conditioned medium will also partially replace the requirement of human lung and skin fibroblasts for FBSP, but in limited preliminary experiments MSA (obtained from Collaborative Research, Waltham, Massachusetts) appears not to be responsible for the activity of the conditioned medium (W. McKeehan, personal communication).

It currently appears that the "decision" by a human diploid fibroblast to enter the cell cycle is influenced by many different components of the culture medium, including lipids, calcium, potassium, and magnesium ions, pyruvate (or other 2-oxocarboxylic acids), the availability of various nutrients, and macromolecular regulatory factors normally supplied from serum. Preliminary growth rate kinetic studies (McKeehan and Ham, 1978a; W. McKeehan, personal communication) suggest that some of the multiplication-promoting small molecules (e.g., lipids, Ca^{2+}, K^+, 2-oxocarboxylic acids) interact very closely with the macromolecular factors from serum, as evidenced by the sharp increase in the amounts of these molecules required for multiplication as the concentration of FBSP is reduced, whereas others (e.g., Mg^{2+} and amino acids) appear to exert their effects at levels more remote from the primary interaction with the macromolecular growth factors.

The apparent complexity of these interactions suggests that many more experiments will have to be performed before the factors and processes responsible for regulating multiplication of normal human fibroblasts are fully understood. However, substantial progress has already been made in reducing the total amount of undefined supplementation needed for satisfactory multiplication, and the amount of FBSP or partially purified factors still needed is small enough so that many new and interesting types of experiments can be undertaken while we are waiting for complete definition of the growth requirements of human diploid fibroblasts.

V. Differences between Skin and Lung Fibroblasts

A. Growth Requirements

In general, we have found the growth requirements of neonatal foreskin and embryonic lung fibroblasts to be quite similar. However, several well-defined differences have emerged:

1. Foreskin fibroblasts are highly dependent on polylysine coating of culture dishes for growth at minimal concentrations of serum protein. Lung fibroblasts also exhibit a positive response to polylysine, but the effect is more variable from strain to strain and generally not as absolute (McKeehan and Ham, 1976).

2. Optimum clonal growth of foreskin fibroblasts can be obtained at pH values slightly more alkaline than those for lung fibroblasts.

3. Foreskin fibroblasts respond to EGF, insulin, and fibroblast growth factor in the presence of small amounts of serum protein, whereas little or no response is seen with lung fibroblasts. A complex pattern of interactions among calcium, pyruvate, and EGF can also be observed with the foreskin fibroblasts, but not with lung fibroblasts, since they do not respond to EGF.

4. Low-temperature trypsinization, which is highly beneficial for growth of both types of fibroblasts with minimal amounts of serum protein, seems to be a more critical requirement for foreskin fibroblasts than for lung fibroblasts (McKeehan, 1977).

B. Other Differences

In addition to these differences in growth responses, we have also observed differences in morphology, particularly at very low concentrations of serum protein. Colonies of foreskin fibroblasts tend to be very flat with virtually no piling of cells on top of one another, whereas lung fibroblasts frequently will pile on top of one another in dense colonies. Also, under conditions where serum protein is very severely limiting, foreskin fibroblasts tend to spread and flatten with very extended cytoplasm and large distances between nuclei, whereas fetal lung fibroblasts are inclined to extend long, thin cytoplasmic processes that are almost axon-like in appearance. These differences in morphology probably reflect differences in adhesion to the substrate, since skin fibroblasts are far more difficult than lung fibroblasts to remove from the culture surface with the low-temperature trypsinization procedure described in Section III, D.

Other investigators have also found differences between skin and lung fibroblasts. Schneider *et al.* (1977) compared various morphological and physiological properties of human fetal lung and fetal skin fibroblasts and concluded that lung cultures had faster cell replication rates, greater tritiated thymidine incorporation into DNA, higher cell numbers at confluency, smaller cell volumes, decreased cellular RNA and protein content, and lengthened *in vitro* lifespans when compared to fetal skin fibroblast cultures. In addition, they reported that hydrocortisone, which is known to have stimulatory effects on fetal lung fibroblasts, "produced a consistent and substantial inhibition of cell population replication in fetal skin fibroblast cultures."

There are also reports in the literature that skin fibroblasts from different parts of the body may possess different biochemical properties. For example, differences in testosterone metabolism have been reported in skin fibroblasts from genital areas and from other parts of the body (Wilson, 1975; Moore and Wilson, 1976). Thus, when we talk about human diploid fibroblasts, and even when we talk specifically of dermal fibroblasts, we are not necessarily talking about a single homogeneous type of cell. It is therefore important to identify precisely the source of each fibroblast culture studied and to take into account variables such as tissue of origin, body region, age of donor, and population doubling level when making comparisons. Furthermore, some investigators have suggested that, even within a restricted area of skin, there may be hot spots that yield good fibroblast growth, while much less growth is obtained from adjacent sites. The relationship between cultured lung fibroblasts and specific cell types in the lung is also a complex problem, as discussed by Bradley *et al.* (this volume, p. 37).

VI.   Selection For and Against Growth of Fibroblasts

A.   Comparison of Media Designed for Fibroblasts, Keratinocytes, and Prostatic Epithelial Cells

The techniques described above for analyzing the growth requirements of human diploid fibroblasts have recently been utilized in studying two types of human epithelial cells. Lechner and Kaighn have studied the growth requirements of normal human prostatic epithelial cells at the Pasadena Foundation for Medical Research (Lechner *et al.*, 1980), and Peehl has studied the growth requirements of normal human epidermal keratinocytes in our laboratory (Peehl, 1979; Peehl and Ham, 1980).

Although the process of quantitative optimization has not yet been completed for keratinocytes and prostatic cells, both studies have resulted in the formulation of new media that support clonal growth of their respective cell types with relatively small amounts of dialyzed serum as the only undefined supplement. EGF and hydrocortisone are also required as ''defined'' supplements by prostatic cells and hydrocortisone by keratinocytes. Thus, there now exist three different media, each of which has been at least partially optimized for a different type of normal human cell. The compositions of these three media, MCDB 105 for fibroblasts, MCDB 151 for keratinocytes, and PFMR-4 for prostatic epithelial cells, are compared in Table I.

It is quite clear from the table that there are significant differences in the nutrient concentrations and balance relationships that are optimal for each of these three cell types. Some of the differences may be artifacts due to use of preliminary data before the optimization process is completed. However, many of the differences appear to be highly significant.

B.   Selective Media

In the development of media for fibroblasts and keratinocytes, we have reached the point where each of the media is quite selective for growth of its own type of cell. When a primary cell suspension is prepared from foreskin tissue as described in Section II, C and inoculated into medium MCDB 105 plus 1000 μg/ml FBSP, the predominant cell type in the resulting culture is fibroblastic, and upon repeated subculture the epithelial cells are lost entirely. When the same cell suspension is inoculated into medium MCDB 151 with the same amount of FBSP plus 10 μg/ml hydrocortisone, fibroblasts grow very poorly, and keratinocytes predominate in the resulting culture.

Our studies of the basis for this selectivity are still incompelte, but a number of differences between the two media probably contribute to their selective growth-promoting properties. Human fibroblasts are very sensitive to inhibitory effects of excess cysteine (Ham *et al.*, 1977) and would not be expected to grow well with the amount of cysteine in MCDB 151 (although the higher levels of other amino acids in MCDB 151 might partially counteract the inhibition). Among the vitamins, the levels of folate, lipoate, and vitamin B_{12} in MCDB 151 are somewhat too high for fibroblasts, and the level of niacinamide is rather low (McKeehan *et al.*, 1977), but none of these alone would be expected to have a major inhibitory effect in the presence of 1000 μg/ml FBSP. The concentrations of adenine, thymidine, and putrescine in MCDB 151 are also somewhat above the optimum for fibroblasts, but probably not high enough to cause a major inhibition of fibroblastic growth. A high concentration of adenine appears to be quite important for keratinocytes.

The optimum concentrations of calcium for growth of the two types of cells appear to be quite different. The concentration of Ca^{2+} in MCDB 151 is clearly too low for fibroblasts (McKeehan *et al.*, 1977; McKeehan and Ham, 1978a), and preliminary data suggest that the rather high concentration of Ca^{2+} in MCDB 105 is excessive for keratinocytes. The concentrations of magnesium, potassium, and phosphate in MCDB 151 are somewhat low for fibroblasts, and the relatively high concentration of zinc in MCDB 151 could be slightly inhibitory to fibroblasts. Finally, the unusually high level of hydrocortisone that must be added to MCDB 151 for satisfactory keratinocyte growth without a feeder layer or conditioned medium (Peehl and Ham, 1979) is probably somewhat inhibitory to skin fibroblasts (Schneider *et al.*, 1977). Supplementation of MCDB 105 with hydrocortisone does not result in satisfactory growth of keratinocytes, suggesting that specific differences between MCDB 151 and MCDB 105, such as the reduced level of calcium, are also of major importance.

When the specific differences in growth responses responsible for the current level of selectivity are fully analyzed, it should be possible to enhance the selectivity further by setting each concentration at the limit tolerated by the desired cell, thereby maximizing the deleterious effect on the cell being selected against. The development of media able to discriminate among various types of cells in a mixed inoculum offers great advantages over current procedures involving differential attachment rates, differential removal of cells, and the use of feeder layers to discourage fibroblast growth.

TABLE I

COMPARISON OF MEDIA FOR NORMAL HUMAN CELLS[a,b]

Component	Fibroblast, MCDB 105	Keratinocyte, MCDB 151	Prostratic cell, PFMR-4
Essential amino acids			
Arginine	1.0 E−3	1.0 E−3	2.0 E−3
Cysteine	5.0 E−5	2.4 E−4	
Half-cysteine			3.0 E−4
Glutamine	2.5 E−3	6.0 E−3	2.0 E−3
Histidine	1.0 E−4	8.0 E−5	2.0 E−4
Isoleucine	3.0 E−5	1.5 E−5	6.0 E−5
Leucine	1.0 E−4	5.0 E−4	2.0 E−4
Lysine	2.0 E−4	1.0 E−4	4.0 E−4
Methionine	3.0 E−5	3.0 E−5	6.0 E−5
Phenylalanine	3.0 E−5	3.0 E−5	6.0 E−5
Threonine	1.0 E−4	1.0 E−4	2.0 E−4
Tryptophan	1.0 E−5	1.5 E−5	2.0 E−5
Tyrosine	3.0 E−5	1.5 E−5	6.0 E−5
Valine	1.0 E−4	3.0 E−4	2.0 E−4
Nonessential amino acids			
Alanine	1.0 E−4	1.0 E−4	2.0 E−4
Asparagine	1.0 E−4	1.0 E−4	2.0 E−4
Aspartate	1.0 E−4	3.0 E−5	2.0 E−4
Glutamate	1.0 E−4	1.0 E−4	2.0 E−4
Glycine	1.0 E−4	1.0 E−4	2.0 E−4
Proline	3.0 E−4	3.0 E−4	6.0 E−4
Serine	1.0 E−4	6.0 E−4	2.0 E−4
Vitamins			
Biotin	3.0 E−8	6.0 E−8	3.0 E−7
Folate		1.8 E−6	3.0 E−6
Folinate	1.0 E−9		
Lipoate	1.0 E−8	1.0 E−6	1.0 E−6
Niacinamide	5.0 E−5	3.0 E−7	3.0 E−7
Pantothenate	1.0 E−6	1.0 E−6	1.0 E−6
Pyridoxine	3.0 E−7	3.0 E−7	3.0 E−7
Riboflavin	3.0 E−7	1.0 E−7	1.0 E−7
Thiamin	1.0 E−6	1.0 E−6	1.0 E−6
Vitamin B_{12}	1.0 E−7	3.0 E−7	1.0 E−6
Purines and pyrimidines			
Adenine	1.0 E−5	1.8 E−4	
Hypoxanthine			3.0 E−5
Thymidine	3.0 E−7	3.0 E−6	3.0 E−6
Other organic compounds			
Acetate		3.7 E−3	
Choline	1.0 E−4	1.0 E−4	1.0 E−4
Glucose	4.0 E−3	6.0 E−3	7.0 E−3

(*continued*)

TABLE I (*continued*)

Component	Fibroblast, MCDB 105	Keratinocyte, MCDB 151	Prostratic cell, PFMR-4
i-Inositol	1.0 E−4	1.0 E−4	1.0 E−4
Linoleate	1.8 E−8		
Putrescine	1.0 E−9	1.0 E−6	2.0 E−6
Pyruvate	1.0 E−3	5.0 E−4	2.0 E−3
Major inorganic ions			
Calcium	1.0 E−3	3.0 E−5	9.2 E−4
Magnesium	1.0 E−3	6.0 E−4	6.8 E−4
Potassium	3.0 E−3	1.5 E−3	4.2 E−3
Sodium	1.3 E−1	1.5 E−1	1.3 E−1
Chloride	1.2 E−1	1.3 E−1	1.1 E−1
Phosphate	3.0 E−3	2.0 E−3	1.2 E−3
Sulfate	1.0 E−3	4.5 E−6	1.6 E−4
Trace elements			
Copper	1.0 E−9	1.0 E−8	1.0 E−8
Iron	5.0 E−6	1.5 E−6	3.0 E−6
Manganese	1.0 E−9		1.0 E−9
Molybdenum	7.0 E−9		7.0 E−9
Nickel	5.0 E−10		5.0 E−10
Selenium	3.0 E−8		3.0 E−8
Silicon	5.0 E−7		5.0 E−7
Tin	5.0 E−10		5.0 E−10
Vanadium	5.0 E−9		5.0 E−9
Zinc	5.0 E−7	3.0 E−6	5.0 E−7
Buffers and indicators			
Bicarbonate		1.4 E−2	1.4 E−2
CO_2	2%	5%	2–3%
HEPES	3.0 E−2	2.8 E−2	3.0 E−2
Phenol red	3.3 E−6	3.3 E−6	6.0 E−6
Supplementation for clonal growth			
Hydrocortisone		2.7 E−5	1.0 E−7
EGF (ng/ml)			5.0
FBSP (μg/ml)	500–1000	1000	200–1000

[a] This table compares the nutrient compositions of media for three different types of normal human cells. Medium MCDB 105 is designed for clonal growth of human diploid fibroblasts (McKeehan et al., 1977, 1978), medium MCDB 151 for clonal growth of normal human dermal keratinocytes (Peehl, 1979), and medium PFMR-4 is for clonal growth of normal human prostatic epithelial cells (Lechner et al., 1980). All three have been at least partially optimized for growth of their respective cell types with minimal amounts of serum protein.

[b] All nutrient concentrations (except as otherwise noted) are in moles per liter. An abbreviated notation has been used in which "E" stands for "times 10 to the power. . . ." Thus, 2.5 E−3 means 2.5×10^{-3} moles per liter. All concentrations have been rounded to two significant figures. Values for inorganic ions are summations of contributions from all sources.

VII. Future Studies

The possibilities for future studies related to the growth requirements of human diploid fibroblasts and other normal human cells are almost unlimited. From our perspective, one of the most urgent tasks is to complete the identification of all extracellular requirements, both qualitative and quantitative, for the multiplication of normal fibroblasts under highly defined conditions. Closely related to this is an analysis of the regulatory mechanisms and interactions, both extracellular and intracellular, involved in determining when a normal diploid fibroblast enters the cell cycle. Obvious secondary studies that should be undertaken as soon as this knowledge is available include differences in growth responses between normal and transformed cells, the mechanisms responsible for density-dependent inhibition of multiplication in culture, the mechanisms responsible for maintaining fibroblasts in an essentially nonmultiplying state *in vivo,* and the mechanisms responsible for the loss of proliferative capacity by senescent fibroblasts in culture.

A second major area that deserves further study involves differences in growth requirements between different types of cells. The development of selective media for different types of cells from the same inoculum will open many new possibilities for the study of cellular differentiation in culture. The improvements achieved in clonal growth of normal diploid cells suggest the very real possibility of using direct cloning from primary cell suspensions to analyze for the presence of different cell types (we already can achieve a plating efficiency of about 1% for foreskin fibroblasts in primary culture). The same understanding of cellular growth requirements that permits the design of selective media should also make possible the formulation of "compromise" media that will support reasonably good clonal growth of highly diverse cell types from a single inoculum.

The availability of defined media that support good multiplication of diverse types of normal human cells would lead quite naturally to studies of the mechanisms of cellular differentiation. The first phase of such studies is likely to be identification of the environmental factors that control the expression of differentiated properties in culture and an analysis of their mechanisms of action. Beyond that, however, there are more fundamental questions concerning the regulatory signals that direct primitive embryonic cells along appropriate pathways of differentiation to become fibroblasts, chondrocytes, myoblasts, adipocytes, keratinocytes, etc. There are also fascinating possibilities of engineering the interconversion from one type of differentiated cell to another *in vitro.*

A critical necessity in all such studies is to achieve a detailed understanding of the environmental conditions required both for proliferation and for differentiation of the cells involved. The technology needed for such studies is available, and the feasibility of the experimental approach has already been proved in studies with human fibroblasts, keratinocytes, and prostatic cells, and with cells from several other species. The only major obstacle that remains is the fact that each type of cell from each species appears to have its own unique set of qualitative and quantitative requirements for multiplication and for differentiation. This means that a major investment of effort by many different investigators will be needed to achieve a full understanding of the requirements of the many diverse types of cells of potential research interest.

ACKNOWLEDGMENTS

I thank Drs. Wallace L. McKeehan and Donna M. Peehl for major contributions to the concepts and data presented in this chapter, Drs. John Lechner and M. Edward Kaighn for providing unpublished data on the growth requirements of prostatic epithelial cells, Kerstin McKeehan, Dennis Genereux, Kathleen Bowen, Kathy Malhotra, Susan Hammond, Stephen Mease, and Linda Miller for excellent technical assistance in the analysis of growth requirements of normal human fibroblasts, and Karen Brown for preparation of the manuscript.

REFERENCES

Chain, B. E., and Knowles, B. R. (1976). *J. Lipid Res.* **17**, 176–181.
Cohen, S., Carpenter, G., and Lembach, K. J. (1975). *Adv. Metab. Disord.* **8**, 265–284.
Dulak, N. C., and Temin, H. M. (1973). *J. Cell. Physiol.* **81**, 161–170.
Goetz, I. E. (1975). *Tissue Cult. Assoc. Man.* **1**, 13–15.
Gospodarowicz, D., and Moran, J. S. (1976). *Annu. Rev. Biochem.* **45**, 531–558.
Greene, A. E. (1973). *In* "Tissue Culture: Methods and Applications" (P. F. Kruse, Jr. and M. K. Patterson, Jr., eds.), pp. 69–72. Academic Press, New York.
Ham, R. G. (1965). *Proc. Natl. Acad. Sci. U.S.A.* **53**, 288–293.
Ham, R. G. (1972). *Methods Cell Physiol.* **5**, 37–74.
Ham, R. G. (1974). *In Vitro* **10**, 119–129.
Ham, R. G., and McKeehan, W. L. (1978a). *In Vitro* **14**, 11–22.
Ham, R. G., and McKeehan, W. L. (1978b). *In* "Nutritional Requirements of Cultured Cells" (H. Katsuta, ed.), pp. 63–115. Univ. Park Press, Baltimore, Maryland.
Ham, R. G., and McKeehan, W. L. (1979). *In* "Methods in Enzymology" (W. B. Jakoby and I. H. Pastan, eds.), Vol. 58, pp. 44–93. Academic Press, New York.
Ham, R. G., Hammond, S. L., and Miller, L. L. (1977). *In Vitro* **13**, 1–10.
Heckman, C. A., Vroman, L., and Pitlick, A. (1977). *Tissue & Cell* **9**, 317–334.
Howard, B. V., de la Llera, M., and Howard, W. J. (1976). *Proc. Soc. Exp. Biol. Med.* **153**, 280–283.
Lechner, J., Babcock, M. S., Marnell, M., Shandar Narayan, K., and Kaighn, M. E. (1980). *Methods Cell Biol.* **21B**, 195–225.
McKeehan, W. L. (1977). *Cell Biol. Int. Rep.* **1**, 335–343.
McKeehan, W. L., and Ham, R. G. (1976). *J. Cell Biol.* **71**, 727–734.
McKeehan, W. L., and Ham, R. G. (1978a). *Nature (London)* **275**, 756–758.
McKeehan, W. L., and Ham, R. G. (1978b). *In Vitro* **14**, 353 (Abstr. No. 77).

McKeehan, W. L., Hamilton, W. G., and Ham, R. G. (1976). *Proc. Natl. Acad. Sci. U.S.A.* **73**, 2023–2027.

McKeehan, W. L., McKeehan, K. A., Hammond, S. L., and Ham, R. G. (1977). *In Vitro* **13**, 399–416.

McKeehan, W. L., Genereux, D. P., and Ham, R. G. (1978). *Biochem. Biophys. Res. Commun.* **80**, 1013–1021.

Martin, G. M. (1973). *In* "Tissue Culture: Methods and Applications" (P. F. Kruse, Jr. and M. K. Patterson, Jr., eds.), pp. 39–43. Academic Press, New York.

Martin, G. M., Sprague, C. A., and Epstein, C. J. (1970). *Lab Invest.* **23**, 86–92.

Moore, R. J., and Wilson, J. D. (1976). *J. Biol. Chem.* **251**, 5895–5900.

Peehl, D. M. (1979). Ph.D. Thesis, University of Colorado, Boulder.

Peehl, D. M., and Ham, R. G. (1980). *In Vitro* (in press).

Pious, D. A., Hamburger, R. N. , and Mills, S. E. (1964). *Exp. Cell Res.* **33**, 495–507.

Poste, G., Papahadjopoulos, D., and Vail, W. J. (1976). *Methods Cell Biol.* **14**, 33–71.

Rheinwald, J. G., and Green, H. (1975). *Cell* **6**, 331–344.

Schneider, E. L., and Mitsui, Y. (1976). *Proc. Natl. Acad. Sci. U.S.A.* **73**, 3584–3588.

Schneider, E. L., Mitsui, Y., Au, K. S., and Shorr, S. S. (1977). *Exp. Cell Res.* **108**, 1–6.

Simpson, W. F., and Stulberg, C. S. (1975). *Tissue Cult. Assoc. Man.* **1**, 211–213.

Smith, J. R., Pereira-Smith, O. M., and Schneider, E. L. (1978). *Proc. Natl. Acad. Sci. U.S.A.* **75**, 1353–1356.

Wilson, J. D. (1975). *J. Biol. Chem.* **250**, 3498–3504.

METHODS IN CELL BIOLOGY, VOLUME 21A

Chapter 17

Prospects for Growing Normal Human Melanocytes in Vitro

SIDNEY N. KLAUS

Dermatology Service,
VA Medical Center,
West Haven, Connecticut

I. Introduction

Researchers have not yet succeeded in growing large numbers of normal human melanocytes *in vitro*. Today, when standard explant or trypsin methods are used, pigment cell cultures become contaminated with large numbers of cells of other types, predominantly keratinocytes and fibroblasts. Generally, in such culture preparations, melanocytes survive for only short periods of time (5–6 weeks).

Two biological characteristics of the human pigment cell system are known to affect adversely the growth of normal melanocytes in an artificial environment. First, melanocytes *in vivo* are not arranged in homogeneous clusters; rather they are scattered individually throughout the entire epidermal sheet. This distribution pattern interferes with the segregation and plating of homogeneous populations of pigment cells. Second, the turnover rate of melanocytes is low, especially when compared with those of the competing populations of keratinocytes and fibroblasts. Both these characteristics lead to the overgrowth of melanocytes by other cell types and make their passage especially difficult.

Although the development of reliable culture procedures for human melanocytes has been stymied, it can be expected that data from recent rather successful studies on guinea pig melanocytes *in vitro* will be usefully applied to the human cell system.

In this chapter current methods for the culture of normal melanocytes will be presented and prospects for the long-term sustained growth of these cells will be examined.

II. Materials and Methods

A. Background

1. HUMAN MELANOCYTES FROM SKIN

Human melanocytes were noted in cultures of skin more than 30 years ago (see Fig. 3; Lewis *et al.,* 1949), and they have continued to be observed in skin cultures since that time (Ingemansson-Nordqvist *et al.,* 1967; Kitano, 1970; Cooper and Cowan, 1974; Hentzer and Kobayasi, 1978). The first study, however, specifically directed toward the culture of human pigment cells of skin was carried out by Hu *et al.* (1957). These workers used an explant technique to grow pigment cells, epithelial cells, and fibroblasts from fragments of skin embedded in chick plasma clots in roller tubes. The morphological appearance and growth characteristics of these cells were examined. It was discovered that the melanocytes *in vitro* maintained their usual dendritic shape and melanin content during the early stages of culture but, as the preparations aged, the number of dendrites decreased and the cells assumed a bipolar configuration. The pigment cells were noted to move out from the explants less rapidly than the adjacent keratinocytes. No pigment cells were seen in mitosis. After several weeks the fibroblasts overgrew the cultures, and the melanocytes disappeared.

A major advance in melanocyte culture was the development by Cruickshank *et al.* (1960) of a method for growing keratinocytes and melanocytes free of fibroblasts in monolayers. With this technique skin fragments were digested with trypsin and the epidermis was separated from the underlying dermis. The epidermal sheets were then disrupted, and the cells were plated on a glass or plastic surface. Cruickshank pioneered the use of small plastic chambers for skin culture, which allowed careful monitoring of cell activity under phase-contrast illumination. Although most of the data in this study concerned guinea pig cells, human keratinocytes and melanocytes were reported to be grown successfully using the same techniques.

Prunieras (1965) also used a trypsinization technique in a study on the morphology and growth characteristics of guinea pig, mouse, and human skin *in vitro*. He reported that the melanocytes disappeared from human skin cultures after 2 weeks.

Kitano and Hata (1972) introduced a modification of Cruickshank's method that used a more elaborate cell dispersion technique, consisting of EDTA, trypsin, and mechanical disruption. Kitano (1976a,b,c) successfully applied this technique in studying the effects of melanocyte-stimulating hormone (MSH), cyclic nucleotides, and theophylline on the growth and structure of human pigment cells.

Serri *et al.* (1973) also grew human pigment cells for short periods of time in plastic chambers. They used the cultured cells to examine the action of the carcinogen 9,12 dimethylbenzanthracene (DMBA) on the human pigmentary system.

2. HUMAN MELANOCYTES FROM CHOROID

Normal human pigment cells from choroid of the eye were cultured by Albert *et al.* (1972). They used a plasma clot technique and succeeded in subculturing pigment cells, although after five passages (12 weeks) fibroblast-like cells became predominant and no melanocytes could be found.

3. EMBRYONIC HUMAN MELANOCYTES

Giovanella *et al.* (1976) reported experiments on the growth of pigment cells obtained from the choroid of human embryos (6–12 weeks' gestation age). The choroid fragments were first minced and then transferred to flasks containing medium. Melanocytes were isolated from the mixed cell preparations and replated. The enormous capacity for growth of the embryonic materials is indicated by the fact that the pigment cell strains were successfully passed up to 27 times.

4. GUINEA PIG MELANOCYTES

Beginning with Cruickshank's successful application of a trypsin dispersion technique for growing cells from guinea pig skin, several investigators have used this model to study the mammalian pigment system *in vitro*. For unknown reasons, guinea pig melanocytes are more readily cultured than human cells. Guinea pig melanocytes tend to divide regularly in culture even though the proliferative rate of the same cells *in vivo* is low. Several authors have focused their attention on the proliferative activity of guinea pig cell cultures. Cohen and Szabo (1968) recorded as many as 26 consecutive cell divisions in melanocytes and found that the daughter cells that resulted from these divisions continued to function normally. Riley (1975) studied the relationship between the proliferative rate of pigment cells and their dendritic contacts. He found that melanocytes underwent division while still maintaining close contact with adjacent pigment cells, although DNA synthesis was rare in melanocytes with multiple contacts. Klaus (1977) examined morphological changes that developed in proliferating melanocytes. He found that dividing cells had shorter dendrites and larger perikaryons than their nondividing counterparts and that their pigmenting capacity decreased with time in culture. He also measured the mitotic cycle of proliferating cells using time-lapse cinemicrography.

B.   Current Methods

Today two standard approaches to the culture of human melanocytes are available. Both methods yield heterogeneous cell populations. One is an explant technique and the other relies on trypsin dispersion. These two techniques will be presented in detail below. (Organ culture, a third method for growing skin, will not be considered because it does not specifically enhance growth of the pigment system.)

1.   EXPLANT TECHNIQUE

a.   Procedure.

1. The area of skin to be cultured is cleansed using 70% alcohol, and thin-shave biopsies are taken using either a knife (no. 11 blade) or a keratome set at 1.5/1000″.

2. The skin samples are placed in a balanced salt solution and cut into small fragments (0.5 mm in diameter) using a sharp knife.

3. The fragments are placed in small clots freshly made from chick plasma and embryo extract (1:1) or, alternately, are placed directly on a glass or plastic surface in the culture vessel. The explants are covered with small amounts of Eagle's minimal essential medium with 10% calf serum to which penicillin (50 μg/ml) and streptomycin (50 μg/ml) have been added.

4. The cultures are maintained in a moisture-controlled incubator in an atmosphere of 5% carbon dioxide in air at 36°C.

5. The medium is changed after 24 hours and three times weekly thereafter.

b.   Results.   The outgrowth of cells from the explants onto the surrounding glass or plastic surface can be seen within 1–3 days (Fig. 1). Melanocytes can be identified under phase-contrast illumination or in stained preparations as dendri-

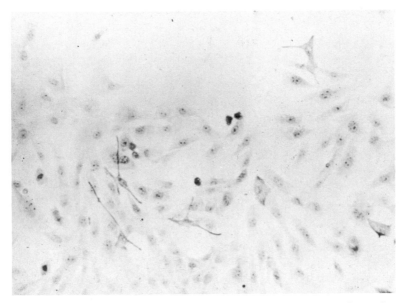

Fig. 1. Dendritic melanocytes scattered among keratinocytes in the outgrowth from guinea pig skin, 12 days in culture. May-Gruenwald–Giemsa stain. Original magnification: ×100.

tic cells containing small pigment granules (Fig. 2). The pigment cells generally move more slowly than either the keratinocytes or fibroblasts and therefore are usually found in the proximal portions of the outgrowth. The major drawback of this technique is that the melanocyte population is sparse, and the cells begin to disappear or degenerate within 2–3 weeks after plating. Thus far attempts to pass these cells using trypsinization have not been successful.

2. DISPERSED CELL TECHNIQUE

a. Procedure.

1. The area of the skin to be cultured is cleansed with 70% alcohol, and shave biopsies are carried out using either a knife (no. 11 blade) or a keratome set at 1.5/1000″.

2. Biopsies are transferred to a petri dish containing a balanced salt solution and are cut into fragments measuring 2–3 mm in diameter.

3. The fragments are floated dermal side down in a watch glass containing a prewarmed solution of 0.25% trypsin (Difco, 1:250) diluted in calcium and magnesium-free Tyrode's solution for 15–30 minutes at 37°C.

4. The dermis is peeled from the grafts with forceps, and the intact epidermal sheets are transferred to a second watch glass containing a small amount of

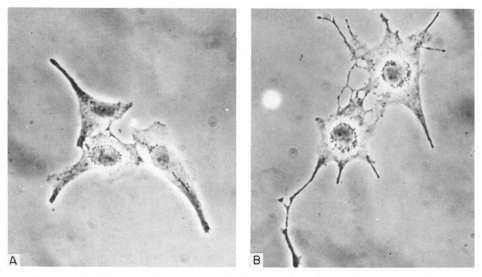

FIG. 2.   (A) Melanocytes from human skin, 15 hours after plating. Phase-contrast. Original magnification: ×400. (b) Human melanocytes with more elaborate dendritic processes, 3 days after plating. Phase-contrast. Original magnification: ×400.

Eagle's minimal essential medium with 10% calf serum to which penicillin (50 μg/ml) and streptomycin (50 μg/ml) have been added.

5. The sheets are placed keratin side down, and the cells from the basalar side of the grafts are dispersed into the medium with paired needles.

6. After all the grafts have been treated, an aliquot of the cell suspension is removed and its cell concentration determined, and sufficient medium (Eagle's minimal essential medium) is added to produce a suspension of approximately 250,000 cells/ml.

7. The cell suspension is added to culture flasks, dishes, or chambers. The cultures are maintained in a moisturized incubator in a 5% carbon dioxide in air environment at 36°C. (If chambers are used, they are incubated coverslip side down for 24 hours to allow the cells to adhere to the underside of the coverslips. After 24 hours the chambers are righted.)

8. Medium is changed after 24 hours and three times weekly thereafter.

9. Once the cells have reached confluence, passage can be attempted by withdrawing the medium and adding a solution of 0.25% trypsin in calcium and magnesium-free Tyrode's solution. The preparations are monitored under a microscope. When approximately one-half of the cells have floated free from the surface (usually 20 minutes), cells are withdrawn and centrifuged for 5 minutes at 500 rpm, washed in medium, and resuspended in medium (100,000 cells/ml). The cells are then replated in new chambers or flasks.

10. Medium may also be added to the original flasks to allow the parent cultures to continue.

b.   Results.   In monolayer preparations melanocytes may be identified by their dendritic shape and melanin content soon after they attach to the glass surface, often within an hour of plating (see Fig. 3). The melanocytes seem to move slowly over the surface of the culture vessel. With this technique the proportion of melanocytes to keratinocytes is slightly higher than when the explant technique is used. In our laboratory mitotic activity has not been noted among human melanocytes, and attempts to pass these cells have not been successful. The cells remain viable for 2–3 weeks.

3.   TESTS OF CELL ORIGIN

Because both the explant and dispersion techniques encourage the growth of heterogeneous cell populations, the positive identification of pigment cells within the cultures assumes importance. During the early stages of culture one can usually rely on the appearance of the cells using the light microscope. However, as the cultures mature, the appearance of the cells often modulates. In our laboratory the origin of cell type is determined by electron microscopic examina-

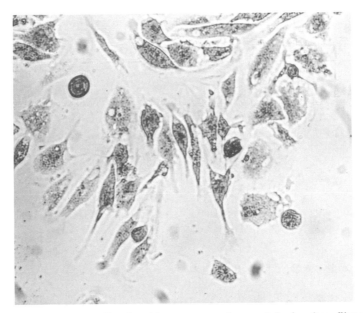

FIG. 3.   Human pigment cells cultured from a compound nevus, 1 day in culture. Phase-contrast. Original magnification: ×100.

tion. Cells to be tested are selected and fixed and embedded *in situ*. We have found that guinea pig melanocytes continue to show characteristic ultrastructure after more than 6 months in culture.

Thus far, the use of histochemical techniques, such as the dopa stain for melanocytes, for identifying cells in culture, has not proved to be reliable.

Another important question concerns the "normalcy" of cells after prolonged culture. Recently we developed a test of normalcy that involves the innoculation of cultured cells back into autologous hosts. In the guinea pig system, cultured melanocytes are harvested using trypsin and centrifuged for 5 minutes at 500 rpm; they are then resuspended in a balanced salt solution at a concentration of 100,000 cells/ml. Ammonium hydroxide blisters are induced on the skin of autologous host animals in non-melanocyte-bearing areas, and 1/10 ml of the cell suspension is injected into the blisters. The development of pigmented spots, at the site of the injections, that contain normally functioning epidermal melanin units confirms the normalcy of the cultured pigment cells.

III.   Perspectives

Short-term mixed melanocyte–keratinocyte cultures have proved useful in studies dealing with the morphological characteristics of these cells, with their interactions, and with their responses to drugs, etc. However, human melanocytes cannot be grown with sufficient homogeneity or in large enough numbers to be used in a wide range of important biological experiments concerning the control of cell turnover, the control of metabolic activity, the nature and identification of membrane receptors, the induction of malignant transformation, etc. Pigment investigators have recently ranked the development of a technique for growing melanocytes as the most important research goal in their field.

Two criteria must be met to ensure the success of a technique for growing melanocytes. First, the cells to be cultured should be relatively homogeneous; second, they should be mitotically active, or at least prepared to divide. In this section possible methods for promoting melanocyte segregation and melanocyte proliferation will be considered based in part on data obtained from guinea pig cultures.

A.   Melanocyte Segregation

As noted above, melanocytes are not grouped in homogeneous clusters *in vivo;* rather they are scattered separately along the dermal-epidermal junction. In most regions of the body pigment cells make up no more than 1 in 8 or 1 in 10 of the total basalar cell population. (The dissociation of melanocytes from their adjacent cells may be especially difficult because the cells are often firmly attached to their neighbors by long dendritic processes *in vivo*. Prolonged enzymatic or mechanical treatment is often needed to separate the cells.)

Several approaches to concentrating the pool of melanocytes prior to culture are now available. Prunieras *et al.* (1976) reported that he could isolate almost pure cultures of melanocytes using the following technique: The dermis is split from the epidermis with trypsin, and the cells at the basalar portion of the epidermis are mechanically dispersed and centrifuged. The cells are washed with a balanced salt solution and then suspended for 3 minutes in a solution of 0.9% sodium chloride containing 0.8% sodium citrate. (This treatment is thought to decrease the attachment of the cells to glass or plastic.) An equal amount of culture medium is then added, and the cells are counted, resuspended in medium in a concentration of 500,000 cells/ml, and replated. Cells that are not firmly attached to the culture surface after 24 hours are removed with the first medium change. Prunieras found that, although most of the cells plated were keratinocytes, the majority of cells that remained behind after the first medium change were melanocytes. In 3 or 4 days, only rare, isolated keratinocytes could be seen in these preparations (see Prunieras *et al.*, 1976, for details). We have had some success in using this procedure in isolating guinea pig melanocytes, but we have not found it useful in dealing with human malanocytes.

A more homogeneous population of melanocytes might also be developed through the physical segregation of cells before they are plated. One method to be considered for separating melanocytes from keratinocytes involves the use of centrifugal elutriation. The elutriator (Beckman Instruments) separates cells on the basis of their size and density. This procedure can be accomplished in less than 1 hour, and the cells can be recovered with almost 100% viability. The elutriator has been used extensively to date to separate cells into the various phases of their mitotic cycle because of the size differences that occur during their progression from mitosis to G_2. It seems reasonable that melanized cells, which are denser and generally larger than keratinocytes, could be isolated in almost pure form by this technique.

A still different approach for ensuring a more homogeneous pigment cell population *in vitro* is to select as a source for culture a cluster of cells with an especially high concentration of melanocytes. In our laboratory we have established pigment cell cultures from nevi and lentigines, which are pigmented skin lesions having a higher melanocyte–keratinocyte ration than normal skin (see Fig. 4). In our studies the cells from these sources did not divide and survived no longer than normal cells (Fig. 4). A drawback to this technique, of course, even if it proves successful, is that there is no assurance that the cells from these lesions are perfectly normal.

Another approach that might be useful in the future involves the use of an agent or drug that selectively favors melanocyte growth. Riley (1970) has reported that gassing mixed melanocyte–keratinocyte suspensions with 95% oxygen and 5% carbon dioxide leads to almost pure melanocyte cultures because the pigment cells show a greater resistance to high oxygen tensions than epidermal cells. Although at present selective melanocidal compounds are known, no selective antikeratinocyte agents have been identified.

B. Melanocyte Proliferation

The second major feature of normal pigment cells that has interfered with their *in vitro* cultivation is their low mitotic rate, especially when compared with those of competing keratinocytes and fibroblasts. (A low mitotic rate is such a constant feature of normal melanocytes that for years many investigators doubted that postembryonic melanocytes had the capacity to divide at all.) Jimbow *et al.* (1975) demonstrated conclusively using electron microscopy that adult human melanocytes underwent cell division in nonstimulated, unexposed areas of the body and that the rate of division increased after exposure to ultraviolet light.

Data from experiments using guinea pig cells indicate that most melanocytes are blocked in G_0 and that the successful growth of melanocytes *in vitro* probably rests on the elimination of this block which prevents the cells from moving into the mitotic cycle. Several workers have examined the effect of exogenous agents on cell turnover. In a recent detailed analysis of the growth requirements of a line of Cloudman melanoma cells, Pawelek (1976, 1979) suggested that cyclic nucleotides tended to decrease the rate of cell turnover in certain strains of melanoma cells while in others it stimulated turnover. Pawelek linked the cyclic AMP in the cells to a cyclic-AMP-dependent protein kinase. A similar control system may be operative in human melanocytes, and changes in cyclic AMP may significantly effect turnover. Kitano (1976b) already has demonstrated that dibutyryl cyclic AMP stimulates human melanocyte proliferation *in vitro,* but that sodium butyrate 5′-AMP and cyclic GMP do not.

It should be kept in mind that other intrinsic factors may also affect the growth of melanocytes. For example, nerve growth factor (NGF) may successfully stimulate pigment cell turnover *in vitro.* Riley (1975) found that contact inhibition in cultures of guinea pig melanocytes tended to decrease the rate of their mitotic activity. Lipkin *et al.* (1977) identified a diffusible factor in hamster melanoma cell cultures, produced by the pigment cells themselves, that caused a reduction in cell turnover when the cells reached confluence.

It also might be feasible to stimulate melanocytes *in vivo* before culture. For example, it has been shown that ultraviolet light will increase melanocyte turnover in skin (Rosdahl, 1978a; Rosdahl and Szabo, 1978). Ultraviolet irradiation of skin 24 hours prior to culture might be used to activate division among cells that otherwise would be mitotically at rest.

Finally, one set of factors that has received surprisingly little attention concerns the effect on pigment cell growth rate of the conditions under which the cells are cultured. Variations in medium formulation, pH, temperature, oxygen concentration, etc., may also have some effect on the rate of proliferative activity. For example, Karasek (1966) noted that variations in keratinocyte growth curves could be related to changes in oxygen tension, serum concentration, and pH.

Experiments should be carried out to measure the influence of these and other factors such as the nature of the substrate, the presence of conditioned medium, the concentration of hormones and vitamins, and the presence of light during culture on the rate of pigment cell growth.

IV. Summary

The current status of the growth of normal human melanocytes *in vitro* has been reviewed. Currently two techniques are available by which human pigment cells can be cultured for short periods of time. One involves the use of an explant, and the other the use of dispersed cells. In both types of preparations the cells ordinarily do not proliferate, and they either disappear or are overgrown by other cell types within a few weeks.

The prospects for growing large numbers of homogeneous normal human melanocytes for long periods have been examined, and approaches have been suggested that may be usefully applied to human cell preparations. The development of a reliable method for the culture of normal human melanocytes has become a top priority in the field of pigment research.

REFERENCES

Albert, D. M., Tso, M. O., and Rabson, A. S. (1972). In vitro growth of pure cultures of retinal pigment epithelium. *Arch. Opthalmol.* **88,** 63–69.

Cohen, J., and Szabo, G. (1968). Study of pigment donation in vitro. *Exp. Cell Res.* **50,** 418–434.

Cooper, J. R., and Cowan, M. A. (1974). Monolayer culture of cells from psoriatic lesions. *Br. J. Dermatol.* **91,** 275–280.

Cruickshank, C. N. D., Cooper, J. R., and Hooper, C. (1960). The cultivation of cells from adult epidermis. *J. Invest. Dermatol.* **34,** 339–342.

Giovanella, B. C., Stehlin, J. S., Santamaria, C., Yim, S. O., Morgan, A. C., Williams, L. J., Leibovitz, A., Fialkow, P. J., and Mumford, D. M. (1976). Human neoplastic and normal cells in tissue culture. I. Cell lines derived from malignant melanomas and normal melanocytes. *J. Natl. Cancer Inst.* **56,** 1131–1142.

Hentzer, B., and Kobayasi, T. (1978). Enzymatic liberation of viable cells of human skin. *Acta Derm.-Venereol.* **58,** 197–202.

Hu, F., Staricco, R. J., Pinkus, H., and Fosnaugh, R. P. (1957). Human melanocytes in tissue culture. *J. Invest. Dermatol.* **28,** 15–32.

Ingemansson-Nordqvist, B., Kiistala, U., and Rorsman, H. (1967). Culture of adult human epidermal cells obtained from roofs of suction blisters. *Acta Derm.-Venereol.* **47,** 237–240.

Jimbow, K., Roth, S. I., Fitzpatrick, T. B., and Szabo, G. (1975). Mitotic activity in non-neoplastic melanocytes in vivo as determined by histochemical, autoradiographic, and electron microscope studies. *J. Cell Biol.* **66,** 663–670.

Karasek, M. A. (1966). In vitro culture of human skin epithelial cells. *J. Invest. Dermatol.* **47,** 533–540.

Kitano, Y. (1970). In vitro cultivation of cells from human epidermis. *Med. J. Osaka Univ.* **21,** 47–52.

Kitano, Y. (1976a). Effects of melanocyte stimulating hormone and theophylline on human melanocytes in vitro. *Arch. Dermatol. Res.* **255,** 163–168.

Kitano, Y. (1976b). Stimulation by melanocyte stimulating hormone and dibutyryl adenosine 3′,5′-cyclic monophosphate of DNA synthesis in human melanocytes in vitro. *Arch. Dermatol. Res.* **257,** 47–52.

Kitano, Y. (1976c). Effects of dibutyryl adenosine 3′,5′-cyclic monophosphate on human melanocytes in vitro. *Acta Derm.-Venereol.* **56,** 223–228.

Kitano, Y., and Hata, S. (1972). The human epidermal cells infected with herpes simplex virus in vitro. *Arch. Dermatol. Forsch.* **245,** 203–210.

Klaus, S. N. (1977). Biologic characteristics of proliferating guinea pig melanocytes. *Yale J. Biol. Med.* **50,** 564.

Lewis, S. R., Pomerat, C. M., and Ezell, D. (1949). Human epidermal cells observed in tissue culture with phase-contrast microscopy. *Anat. Rec.* **104,** 487–503.

Lipkin, G., Knecht, M. E., and Rosenberg, M. (1977). Role of a diffusible factor in the in vitro growth control of malignant melanocytes. *Yale J. Biol. Med.* **50,** 564.

Pawelek, J. P. (1976). Factors regulating growth and pigmentation of melanoma cells. *J. Invest. Dermatol.* **66,** 201–209.

Pawelek, J. P. (1979). Evidence suggesting that a cyclic-AMP-dependent protein kinase is a positive regulator of proliferation in Cloudman S91 melanoma cells. *J. Cell. Physiol.* **98,** 619–626.

Prunieras, M. (1965). "Culture de l'épiderme de mammifère adulte," pp. 55–56. S.P.E.I., Paris.

Prunieras, M., Moreno, G., Dosso, Y., and Vinzens, F. (1976). Studies on guinea pig skin cell cultures. *Acta Derm.-Venereol.* **56,** 1–9.

Riley, P. A. (1970). Studies of normal melanocytes in culture. *J. Invest. Dermatol.* **54,** 96.

Riley, P. A. (1975). Growth inhibition in normal mammalian melanocytes in vitro. *Br. J. Dermatol.* **92,** 291–304.

Rosdahl, I. K. (1978a). Melanocyte mitosis in UVB-irradiated mouse skin. *Acta Derm.-Venereol.* **58,** 217–221.

Rosdahl, I. K. (1978b). Mitotic activity of epidermal melanocytes in UV-irradiated mouse skin. *J. Invest. Dermatol.* **70,** 143–148.

Serri, F., Pisanu, G., and Cantu, G. (1973). Human melanocytes, keratinocytes and carcinogens in vitro. *G. Ital. Dermatol.* **108,** 73–82.

Chapter 18

Integumentary System—An Overview

STUART H. YUSPA

In Vitro Pathogenesis Section,
Laboratory of Experimental Pathology,
National Cancer Institute,
Bethesda, Maryland

The study of mammalian skin provides a unique opportunity to enhance our understanding of a variety of biological processes. Maintenance of normal architecture and function in the skin requires specialization of multipotential stem cells, cell–cell communication, epithelium–mesenchyme interaction, and a variety of intrinsic and extrinsic regulators of cell growth and differentiation. There are practical reasons for choosing skin to study these processes. Skin is the most abundant and accessible tissue in the organism. It appears to maintain its integrity for some time after removal from the host or after the death of the host. *In situ* it can be subjected to a number of pharmacological manipulations without endangering the whole organism. Combinations of *in vivo* and *in vitro* experimentation can be performed with facility. A variety of genetic or acquired conditions have been described in humans and other species in which one or more of the normal skin functions are altered, providing experimental models for studies on these functions. Genetic defects in a number of diseases were first delineated by studies on skin cells (e.g., xeroderma pigmentosum, diseases of purine metabolism). Skin is also a tissue that has a high rate of neoplastic transformation in humans and certain rodents and thus is useful as a model for carcinogenesis (Yuspa, 1978).

Recent data indicate that the process of terminal differentiation in epidermis is quite similar across species lines and that major products of terminal differentiation in the epidermis (e.g., keratin proteins, stratum corneum basic protein) have immunological identity among species (Dale *et al.*, 1978; Fuchs and Green, 1978; Lee *et al.*, 1976; Steinert *et al.*, 1979). This allows the application of research tools and the extrapolation of research results between experimental models.

The skin is a complex organ in structure and function. At least four major cell types reside in this tissue, excluding vascular, adipose, and muscular components. The keratinocyte is the major epithelial cell of the skin. This cell or a related stem cell has the capacity to form both the stratified epithelium of the epidermis and the epidermal appendages found largely in the dermis. This is particularly remarkable, since the biochemical composition of the differentiation products of the stratified epidermis and the appendages, including hair, is very different (Seiji and Bernstein, 1977). The major stromal component of skin is the dermal fibroblast. This cell is an important regulator of epidermal function, as well as a provider of stromal support. The regulatory function of dermis appears to be mediated through both humoral factors and direct cell contact (Briggaman and Wheeler, 1968; Elgjo, 1974). Melanocytes are a third major cell in the skin and provide pigment for keratinocytes. Pigmentation protects the epidermis from untraviolet damage. A fourth cell type is the Langerhans cell, found in the midepidermis. Cytoplasmic processes of these dendritic cells extend between keratinocytes down to the dermis–epidermis junction (Shelley and Juhlin, 1978). While the exact function of this cell has not been elucidated, it appears to belong to the reticuloendothelial system and to be involved in immunological reactions of the skin.

This brief review indicates that present data have provided a basic understanding of the physiology of the skin and its individual cell types. However, exact delineation of the biochemical or genetic regulation of the functions of each cell type will require a great deal more study. Many of these studies will be best performed *in vitro* on isolated cells or reconstituted cell mixtures. This chapter describes the current state of development of *in vitro* systems for three cell types from human skin: keratinocytes, dermal fibroblasts, and melanocytes. With further refinements these models will provide an essential link to understanding the molecular biology of normal skin and the defects associated with both specific skin and systemic diseases.

REFERENCES

Briggaman, R. A., and Wheeler, C. E. (1968). *J. Invest. Dermatol.* **51,** 454–465.
Dale, B. A., Holbrook, K. A., and Steinert, P. M. (1978). *Nature (London)* **276,** 729–731.
Elgjo, K. (1974). *Virchows Arch. B* **16,** 243–247.
Fuchs, E., and Green, H. (1978). *Cell* **15,** 887–897.
Lee, L. D., Baden, H. P., Kubilus, J., and Fleming, B. F. (1976). *J. Invest. Dermatol.* **67,** 521–525.
Seiji, M., and Bernstein, I. A., eds. (1977). "Biochemistry of Cutaneous Epidermal Differentiation." Univ. of Tokyo Press, Tokyo.
Shelley, W. B., and Johlin, L. (1978). *Acta Derm.-Venereol.* **58,** Suppl. 79, 8–22.

Steinert, P. M., Idler, W. W., Poirier, M. C., Katoh, Y., Stoner, G. D., and Yuspa, S. H. (1979). *Biochim. Biophys. Acta* **577,** 11–21.

Yuspa, S. H. (1978). *In* "In Vitro Carcinogenesis, Guide to the Literature, Recent Advances and Laboratory Procedures" (U. Saffiotti and H. Autrup, eds.), pp. 47–56. US Govt. Printing Office, Washington, D.C.

SUBJECT INDEX

293

CONTENTS OF PREVIOUS VOLUMES

(Volumes I–XX edited by David M. Prescott)

Volume I

Volume VIII

Volume X

Volume XI

Volume XIV

Volume XVI

Volume XVII

Volume XX